EXPLORING THE
HISTORIC
SAN JUAN
TRIANGLE

P. David Smith

WESTERN REFLECTIONS PUBLISHING COMPANY®

Lake City, Colorado

ISBN 978-1-937851-46-0

Printed in the United States of America

Cover postcards and author photo courtesy P. David Smith

Western Reflections Publishing Company®
P. O. Box 1149
951 N. Highway 149
Lake City, Colorado 81235
www.westernreflectionspublishing.com
(970) 944-0110

ᗌ Contents ᗍ

The San Juan Mountains are some of the roughest and most spectacular in North America. This is the view looking northeast from Engineer Mountain. Bill Fries III Photo. Author's Collection.

↶ Dedication ↷

The Ouray County Chamber of Commerce originally published the forerunner of this book, *Mountain Mysteries,* in 1981. I had begun the process of collecting information about the history of the San Juans shortly after moving to Ouray in 1976. My wife, Jan, and my children, Tricia, Tami, and Stephen, spent many hours with me in our old 1949 Willys Jeep as we slowly made our way over the local mountains in search of new adventures. I took photographs of more than a thousand mines, cabins, mills, and other man-made structures on these trips. I am very pleased that I still have those photographs because many of the structures no longer exist. In the 1970s I had little idea of the fragility of the solid-looking structures I was looking at, but year by year there is less and less in the way of historic man-made objects left to enjoy in the beautiful San Juan Mountains.

Man himself does most of the damage. Until the early 1940s most of the mines and camps of the San Juans remained completely intact, everything waiting for the men to return, open the mines back up, occupy the cabins again, and once more start the crushing process in the mills. However, during World War I most of the easily accessible steel was salvaged from the mines, railroads, and ghost towns to help with the war effort. Much of the lumber was taken after the war to be used in the construction of motels and other attractions catering to the tourists that were starting to recognize the attractions of the San Juans. Vandals or treasure seekers, along with the wind, took much of what remained; and heavy snows conspired to push many of the remaining structures to the ground. The spring snowmelt continues to slowly wash away the mine dumps. Most of the ghost towns are now no more than mounds of dirt. Many of the newer structures are just piles of rubble. The San Juans are quick to remind us just how small man's achievements really are and how easily they can be erased. But that is one of the purposes of this book—to help us to remember and to preserve the heroic achievements that took place here. This book is, in part, dedicated to those early pioneers.

However, I have also survived three wonderful men who helped greatly with the forerunner of this book. Waldo Butler died in 1980 shortly before *Mountain Mysteries* was first published, and I can still remember the first time I read the opening words that he proposed for the book: "Fire and ice shaped these mountains and created one of the most spectacular landscapes in the world." Waldo helped me to realize that these mountains deserve more than a tour guide to explain what they mean to us. Marvin Gregory helped me put together a revised version of the book in 1984. He helped me recognize the importance of photographs to a book like this, and Marvin was also a stickler for details. Over and over he would repeat that he saw no need for exaggerations or embellishments. He knew that the area has too much exciting, true history to resort to fabricating stories or events. Marvin died in 1992, but I share his passion for writing only true, believable history that can be supported by documented evidence. The third man—Jack Swanson—is also to be credited for this book. After the Ouray Chamber of Commerce decided not to reprint the book, Jack published *Mountain Mysteries* and taught me the qualities that make a good publisher. He encouraged me; he talked with me; he helped me see what interested the reader. Jack had an enormous sense of pride in the books he published. He wanted them to be the absolute best in design and content. Jack's work showed results because *Mountain Mysteries* sold over 30,000 copies. Jack died in 1998 after a brave battle with cancer.

So, I no longer have three wonderful friends to help me write this book. Even though they have been gone just a short time, they would certainly be surprised to see how much our land has changed in the last few years. However I can still feel the hands of those three men on my shoulder as I write. I owe a lot to them, and this world is a lesser place without them.

CAUTION

Although the San Juan Mountains are beautiful, they also can be very dangerous—

Always use common sense when dealing with a situation you don't understand. Conditions might have changed between the time this book was written and the time you are using it. Make sure you have adequate provisions and up-to-date directions. This book is intended to supplement and not to replace the need for a good jeeping or hiking map.

When hiking, always let someone know where you are going and try not to hike alone. Be sure to take adequate provisions, have the proper maps, and don't be afraid to ask the locals for directions. Be careful to let your body adjust to the altitude for a couple of days. Leave plenty of hiking time when going uphill so you can stop and rest often. Remember at 10,000 feet there is only half the oxygen that is available at sea level. Going down the steep slopes can be hard on the knees. It helps to have a hiking staff. Be sure to wear warm clothing and take rain gear. Always carry plenty of water because the high altitude causes your body to dehydrate much quicker than at a low altitude. If you get lost, stop, keep warm and dry, and don't panic. If you can't find your way out, stay put; and if you have told someone where you are going, someone will be looking for you before long. If you have to move, try to go downhill and then downstream. That is where civilization is usually found.

If camping, please make a fire only in a fire pit and only under the conditions in which a fire is allowed. Forest fires are an ever-present danger. Take only photos on your visits and leave only footprints. Remember that mines and cabins are private property. Don't do anything to cause our historic structures to deteriorate any faster than they already are because of the harsh weather. Whether jeeping or camping, please remember to be considerate of those who come after you. Don't leave trash to spoil someone else's adventure. In fact, if you really want to do a good deed, pack out the trash of others.

When driving, remember that snow, sleet, or ice might be on the roads at any time of the year. Clay and mud also make some of the roads very slick. Remember to allow plenty of time for your outing. Many of these roads are traveled at five miles per hour or less. People gauge distance here in hours, not miles! Always stay on your side of the road, no matter how steep the drop-off. Look ahead and not down. Remember that the vehicle going uphill has the right-of-way on one-lane roads, but use a little common sense and courtesy. If you see a good place to pull off nearby, then take it, and let the downhill vehicle go by. Gear down when going downhill so you won't burn out your brakes, and be especially careful when you stop on an incline—block your wheels.

If you have any concerns at all about your driving ability, take a jeep tour; the tour drivers are experienced on the roads, and most of them know a lot about the local history, geology, and fauna. Whether walking, climbing or jeeping, remember that it always looks easier going up than coming down.

These warnings are meant to point out some of the dangers that go along with the pleasures of our awesome country. Nothing can compare to the San Juan Mountains' fresh air, clear streams, beautiful scenery, and nights of sparkling stars. Enjoy yourself in God's Country!

Introduction

It takes a long time to explore the San Juan Mountains. There are people who have spent their entire life in Southwest Colorado and still haven't seen it all. These mountains are rugged! They give up their secrets only after a long fight. So most of the enjoyment has to be fought for—either in a bouncing, twisting four-wheel-drive vehicle or while gasping for breath when hiking on one of the high mountain trails.

Today's jeep roads are generally the old wagon roads of the past, and many of the present hiking trails were pack trails to long abandoned mines. These roads and trails have made many places easily accessible that otherwise would never be seen but by a hardy few. However, even just a drive on paved roads around the San Juan Skyway will give the visitor a chance to sample the real flavor of the area. Silverton, Ouray, Lake City, and Telluride can all be explored from the family car, yet these towns look almost the same as they did in the 1890s.

Sometimes there is no trail, and one must head out on foot cross-country (please don't drive vehicles off the existing trails). If you are willing to spend the time and take the trouble, the reward is well worth the effort, because these are perhaps the most beautiful mountains in the United States—maybe even the world. There is an added bonus to be found while exploring the San Juan Mountains. They also abound with history—an amazing history. There are still relics of the old mining days—abandoned mines, massive machinery, cabins, and mills. Some of the most famous narrow gauge railroads in the world were in the San Juans, and dozens of ghost towns hide in the rugged terrain.

Because most people want to know something about the types and amounts of precious minerals that the San Juans produced, I have also tried to address this topic in this book. However, a few factors need to be considered. Gold and silver values are usually stated as ounces per ton of rock. Other minerals such as lead, copper, and zinc are usually given as percentages of the total ore. Ore is defined as any rock that contains enough minerals to be economically produced. In the San Juans ore is usually contained in veins, which can be just inches thick or up to many feet wide. Within a vein there can be several other veins or "pay streaks." A tunnel needed to be at least three feet wide and five feet high to accommodate the men mining the ore. This requirement meant that a small, rich vein might not be economical because of all the barren rock that had to be taken out to get to it. Also some ore appeared in pockets, while some veins might go great distances horizontally or vertically. Another factor was how easily the ore could be refined or smelted.

Some ore figures are given as values right out of the ground, other figures are for sorted ore, some are projected from a single piece of extremely rich ore, and still others are given for the ore after it has been concentrated (a large percentage of the barren rock taken out). All this makes it almost impossible to compare values from one mine to another, but it does give you some idea of the type of mineral being produced and the richness of the vein.

Another added bonus when exploring in the San Juans is the chance to experience some of the finest sport and recreational activities available in the United States—fishing, hunting, hiking, mountain climbing, rafting, skiing, snowmobiling, ice climbing, mineral collecting, camping, horseback riding, nature watching, and at least a dozen other pursuits. Whatever sport or activity interests you—whether it's a wilderness experience or a guided four-wheel drive tour—it can probably be found in the San Juan Mountains.

Because of their geological complexity, dramatic sculpting by glaciers, and extremely high elevation, the San Juans Mountains have fascinated, awed, and inspired all who have entered this majestic region. The Reverend J. J. Gibbons, an early Catholic priest, wrote in his book *In The San Juans* in 1898:

> No man can travel through the (San Juan) mountains without a deepening impression of The Creator; no one can stand in the presence of the snow-capped peaks, over which sunshine and shadow pursue each other, without feeling an impulse to elevate his soul to God, the author and finisher of the beautiful and the sublime. A trip to the mountains convinces the religious mind of the existence of a divine power, wisdom and goodness, and inspires men of good will with the resolution to seek first the kingdom of God and His justice. Where all is so divine, surely the spirit of man should not be merely human.

At almost any time of spring, summer, and fall there is brilliant color in the foliage of the San Juans. In early spring the lime green of the newly budding trees works its way up the mountains, and then within a few weeks the spring flowers begin blooming in the lower valleys and continue up into the high mountains, producing nature's colorful carpet until as late as August. In the fall the process reverses itself as a riot of reds, oranges, and yellows work their way back down the mountain where they arrive in the valleys in late September. The scrub oak and cottonwoods in the foothills are usually in full color by early October. In the winter it is an absence of color that one notices—whites and cold blues prevail with only the occasional green of the evergreen trees peeking out from their blanket of snow.

Almost every type of life zone is present in the San Juans, because every thousand feet gained in elevation is the equivalent of traveling 350 miles north at sea level. Sunny south-facing slopes have the equivalent plant life of northern slopes or shady places a thousand feet or more higher in altitude.

Located between about 5,000 to 7,500 feet in elevation is the Foothills or Lower Montane zone. It is a dry area—just one step above the desert. Cottonwood, scrub oak, box elder, piñon, sage, and juniper are plentiful, and spring comes to the foothills in April, May or June. The Canadian or Montane zone is found from 7,500 to 9,500 feet. Pine, aspen, and willows are the predominant trees, and flowers are normally in full bloom from June through July. Aspen are considered by some to be the largest plant on earth because they grow from a connected root system. This phenomenon explains why half a hillside of aspen might be an unusual color in the fall or change color simultaneously—all of which adds to the magnificent fall beauty.

From 9,000 feet to timberline (about 11,500 feet) is the Subalpine zone. Spruce and fir predominate. Flowers bloom from late June through August. There is considerable precipitation. Above timberline trees do not grow, and the area is called Alpine or Arctic. Mosses, lichens, and grasses are predominant, but they can be just as beautiful as the flowers and trees if one will only take the time to look closely. Also found at this high elevation is tundra, a very rich and beautiful soil. Tiny flowers are often part of the tundra. Two inches of this soil takes thousands of years for nature to produce, so it is one of the most fragile and delicate forms of life in the San Juans (in fact in the world). So please stay on the traveled paths! It can take hundreds of years for damaged tundra to repair itself.

The climate of the San Juans varies radically. For example, only rarely does the thermometer drop below zero in Ouray in the winter and forty-degree days are common.

The temperature in Silverton (just twenty-five miles away) goes well below zero almost every night in the winter, yet its high temperature is often greater than Ouray. Snow is light and dry, but abundant. Wind is usually nonexistent except during severe storms and sometimes in the spring. Summers are usually cool with highs in the seventies and eighties and lows in the fifties or sixties. Almost daily, rains fall in July and August (during the "Monsoon Season"). The warm air coming off the deserts to the west is pushed up, and as it cools, it drops moisture also coming in from the south. The summer thunderstorms and floods often cause mud or rock slides.

Autumn is usually dry with warm days and chilly nights. The low humidity and lack of wind always make the temperature seem warmer than it actually is. This phenomenon leads many visitors to go into the mountains unprepared. Hypothermia (lowering of the body's temperature) can happen quickly—even in the summer at high altitudes. The wetness and wind of summer thunderstorms can severely lower one's body temperature. Often it will snow even in the middle of the summer at high altitudes, although it melts quickly when the sun comes out. To avoid hypothermia, wear warm clothes, take rain gear, eat snacks, drink water, and try to avoid exhaustion. Another danger of summer thunderstorms is lightning. Stay off high ridges or mountain peaks during a storm, don't huddle together, and stay in your vehicle if you have one.

In the winter one must be wary of avalanches. Be especially careful after heavy snows, travel in groups, and try to carry an avalanche beacon. There have been several hundred avalanche deaths over the last century, and there still are a few casualties every winter. Any time of year the sun can burn you quickly because of the thinness of the air and lack of pollution, but it is especially bad in the winter when you get hit twice because of the sun reflecting off the snow. Snow blindness, because of the bright light, is a common problem, even when driving at night.

To give some idea of how extreme the weather can be, the recorded San Juan record for rain in one day is eight inches (at Gladstone), but there are many years when it has not rained at all during the entire month of June. Sixty-two inches of snow fell in one day (at Purgatory), 144 inches in a week, and 581 inches in one year (both at Savage Basin near Telluride). The most snow on the ground at one time was 184 inches at Animas Forks in 1884 (but there were drifts up to seventy-five feet deep)! Yet many years there is no measurable snow until late December. The average temperatures also vary radically. Some years may be twenty or thirty degrees above or below normal.

Wildlife in the San Juans includes chipmunks (they will eat out of your hand at Box Canyon Park in Ouray), marmots or "whistle pigs" (they look like large golden or brown groundhogs), ptarmigan (a very tame kind of grouse that is white in the winter and brown during the summer), squirrels, beaver (recently making a real resurgence in the San Juans), skunks, porcupines, rabbits, deer, elk, and mountain sheep. Less frequently seen are black bear, mountain lions, bobcats, lynx (they have recently been transplanted and most are wearing radio collars), moose, wolves, and coyotes. Deer and elk are seen mostly during early morning hours and at dusk. In the summer they inhabit the high grassy meadows, but as winter approaches the herds move to lower elevations. Large herds of both deer and elk can commonly be seen alongside (and sometimes in) the roadway during winter. Contrary to some tourists' opinions the deer do not turn into elk in the fall.

When hunting, please be sure that you are on public land or have the permission of the landowner. Pay close attention to weather reports in the fall. The weather can change very quickly! Remember that the road you came in on may be impassible after a heavy snowfall or rain. And look carefully before you shoot!

Many bird species are found in the San Juans, including swallows, magpies, robins, bluebirds, jays, sparrows, eagles (mainly golden, but the bald eagle is also frequently seen during the winter), hawks, grouse, pheasant, quail, doves, ducks, geese, and owls. Many varieties of hummingbirds feed from the abundant flowers or bright feeders throughout the San Juans in the warmer months.

Trout are either native or are stocked in most of the San Juan streams. Varieties of trout in the area include cutthroat, rainbow, brook, and some brown trout. Many of the high altitude lakes have good fishing. Watch for the signs that may place special restrictions on the streams or lakes. Some areas are limited to artificial lures or flies, and other areas are posted. Spring runoff normally makes stream fishing poor and dangerous in May and June because of the deep, murky, and swiftly moving water. At this time of year it is best to fish the high mountain creeks or lakes.

The flowers in the San Juans are overwhelming, perhaps the best in all of Colorado. Remember that "spring" in the high mountains is at the same time as summer at lower elevations. The high mountain flowers are very fragile, so please do not pick them—in fact, it is illegal to pick the state flower, the columbine, as well as many other flowers. A good local flower book will add a lot of enjoyment to your hiking or jeeping.

No variety of poisonous snakes is known to exist above 7,500 feet in the San Juans and only a few garden-types are found. However in May and June, you should look out for the ticks (which can carry Rocky Mountain fever and Lyme disease).

One other unique feature has been added to this book—an alternate way to explore the San Juans and never leave your chair; that is, by exploring the area by reading some of the many books written about the San Juans. There are literally hundreds of books (some written one hundred years ago or more) that explain the history of this area, and I will try to mention some of the best. They can give many hours of pleasant recreation on those cold, wintry nights when the high country is inaccessible.

There are two books that will increase your appreciation of the mining relics that you will find all over the mountains. Beth and Bill Sagstetter's *The Mining Camps Speak* does a great job of describing the "trash" you will find around the mines. It is amazing the story it can tell. Eric Twitty's *Riches to Rust* explains the mining equipment in detail, even down to the missing equipment revealed by concrete foundations.

I frequently refer to three classic authors. One is Ernest Ingersoll, a travel writer and adventurer who first came to the San Juans with the government-sponsored Hayden expedition in 1875 and then returned many times. He incorporated his travels into many books that were published into the late 1880s. Another author was Frank Fossett, who was also a mine promoter of sorts. Fossett was in the San Juans throughout the 1880s and 1890s and again wrote several books that always mentioned the details of the local mines. George Crofutt wrote guidebooks for the railroads about the west. From 1869 to 1892 he sold more than a million and a half books—his Colorado guide being *Crofutt's Grip-Sack Guide to Colorado*, which was first published in 1881. All are such wonderful, detailed writers that they bear rereading; but be aware that they were a little prone to exaggerate, as were most of the writers of the time. These mountains have such an exceptional beauty and history that they easily inspire the writer—yours truly being tempted but I have tried to stick to the facts. Enjoy!

MAP OF THE SAN JUAN TRIANGLE

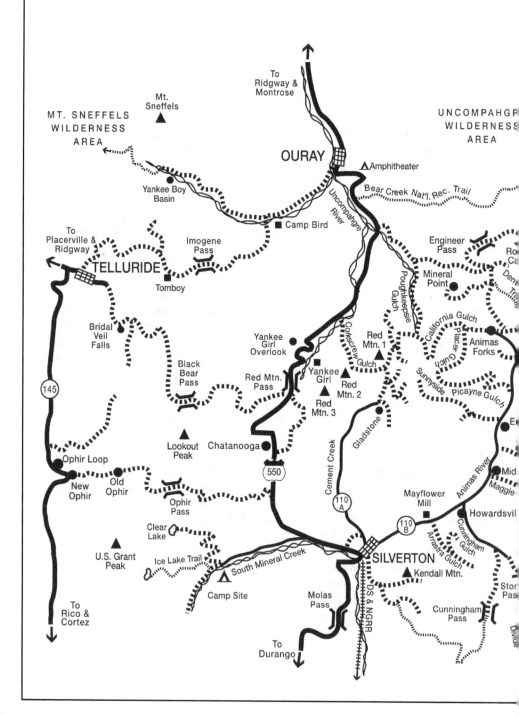

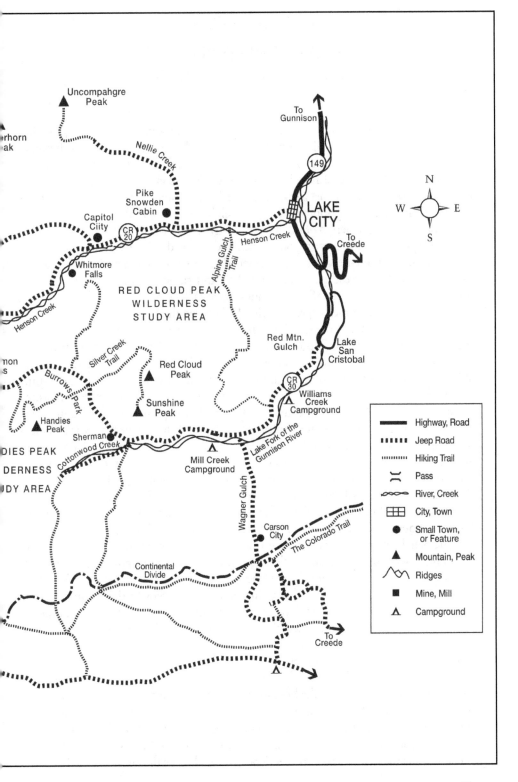

Uncompahgre
Peak

rhorn
ak

Nellie Creek

To
Gunnison

149

Pike
Snowden
Cabin

Capitol
Ciity

CR
20

LAKE
CITY

To
Creede

Henson Creek

Alpine Gulch Trail

Whitmore
Falls

RED CLOUD PEAK
WILDERNESS
STUDY AREA

Henson Creek

Red Mtn.
Gulch

Lake
San
Cristobal

non
s

Silver Creek
Trail

Red Cloud
Peak

CR
30

Williams
Creek
Campground

Burrows Park

Sunshine
Peak

Handies
Peak

Sherman

Lake Fork of the
Gunnison River

DIES PEAK
DERNESS
JDY AREA

Cottonwood Creek

Mill Creek
Campground

Wagner Gulch

Carson
City

The Colorado Trail

Continental
Divide

To
Creede

N
W   E
S

▬▬▬	Highway, Road
▪▪▪▪▪	Jeep Road
....	Hiking Trail
⊱⊰	Pass
∿∿∿	River, Creek
⊞	City, Town
●	Small Town, or Feature
▲	Mountain, Peak
∧∧∧	Ridges
■	Mine, Mill
⛺	Campground

Chapter One

THE EARLY HISTORY OF THE SAN JUAN MOUNTAINS

From the time that men first set foot in the San Juan Mountains, the range has fascinated, awed, and inspired all that have seen them. The San Juans cover ten thousand to twelve thousand square miles of Southwest Colorado, depending on how you define its boundaries. Regardless of how it is defined, the San Juans are the largest mountain range in Colorado (about the size of Rhode Island, Connecticut, and Massachusetts combined and composing approximately one-eighth of Colorado).

Before the volcanic period of thirty to thirty-five million years ago, the San Juan area had been a part of several oceans (you can still find petrified sea shells in the mountains). At that time an immense dome was formed that in the center was about twenty-six thousand feet above sea level. Some eight thousand cubic miles of volcanic ash were thrown into the atmosphere from volcanoes near its center. The dome was cracked and fractured by the immense pressure that had caused the land to rise. This immense increase in altitude caused the Continental Divide to bulge considerably to the west at the point of this activity, and today, in some places, as much as nineteen layers of rock formations are exposed. Frank Fossett in 1879 pointed out the obvious: "There is probably more country standing on edge in this section than anywhere else under the sun."

In other places in the San Juans there were craters (technically called calderas) formed by volcanoes that collapsed upon themselves leaving deep depressions that were anywhere from a hundred feet to a hundred miles across. Most of these craters appear as valleys or bowls in the vicinity of what is now Silverton, Lake City, and Creede. All of this up and down motion made the San Juans very rugged. When the famous historian Hubert Bancroft published his history of Nevada, Colorado and Wyoming in 1890, he declared the San Juans as "the wildest and most inaccessible region in Colorado, if not in all of North America." Even after millions of years of erosion, fourteen San Juan peaks exceed fourteen thousand feet and hundreds more are over thirteen thousand.

The San Juan Mountains were born of fire, but they were mostly shaped by ice. Generally it was not the volcanic action but rather the incessant movement and grinding of glaciers that scoured out the present-day valleys and cirques (or bowls), leaving the sharp needle-shaped peaks and large mountain ranges. Slowly, the mountains wore down at the rate of about one inch per decade. At this rate the area will be flat again in about twelve million years! One exception to the rule occurred when the U.S. Geological Survey refigured the height of U.S. mountains in the year 2002. Most of the San Juan mountains "grew" by four to six feet as a result of more exact measurements.

Water, acid, and gas-based mineral solutions rose up through the cracks and fissures. When they cooled they left valuable minerals locked in hardrock vein deposits. Most of the major mines of the San Juans are found along these faults-although finding and then following a vein is a very complex job. A vein might vary from fabulous richness to extremely low-grade ore within just a few feet. In the San Juan Mountains very few valuable minerals were found in placer deposits, which were the familiar gravel deposits

A jeep pauses for the spectacular view from the top of Imogene Pass. Bill Fries III Photo. Author's Collection.

containing gold nuggets that were panned by the California prospectors in the 1850s. Ironically, there is sometimes decent panning today downstream from old millsites, which often introduced flakes of unrecovered gold and silver into the creeks by inefficient milling processes.

As a general rule the higher up that minerals are found in the San Juans, the richer they tend to be. Many of the most famous San Juan mines were therefore discovered near the very tops of the mountains. In the San Juans, silver is the predominant of the valuable minerals, but there are also rich, isolated deposits of gold and very large areas of base metals such as copper, lead, and zinc. The most frequent ores in the San Juans are galena (lead and silver) and pyrites (also called "fool's gold," but which can contain valuable minerals).

The first human inhabitants of the San Juans were probably Paleo Indians who appeared on the scene about seven thousand to thirteen thousand years ago. The Paleo Indians traveled lightly through this beautiful land. During their time there were still remnants of the great glaciers that had carved out most of the San Juan Mountains. Today, fire pits, rough stone tools, and a few beautiful spear points are about the only trace that can be found of Paleo occupation. They hunted the last of the woolly mammoths and great herds of bison. They were followed by the Archaic Culture (two thousand to seven thousand years ago), then the Anasazi and Fremont Cultures, who arrived about the time of Christ.

These latter groups left behind slightly more evidence of their existence. Especially unique are their cliff dwellings, which are found on the edges of the San Juans,

13

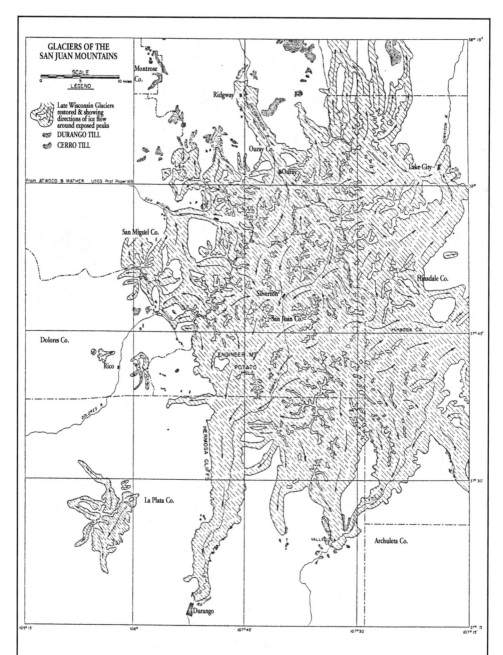

At one time glaciers covered all but the highest peaks in the San Juans. Their power was enormous and the valleys that they carved out created most of the mountains that we now see. Their movement was in the same directions as our rivers now flow - as shown on this complex diagram of the glaciers of the San Juans. Atwood & Mather USGS Professional Paper 166 indicating that they traveled into the heart of these mountains only in the summer, and then only as nomadic wanderers. Anasazi pottery has, however, been found as high as twelve thousand feet in the San Juans, probably brought in by travelers. There were also people who lived near the northwest end of the San Juans, now referred to as the "Uncompahgre Complex." Unfortunately we know little about them, except that they too ventured into the San Juans only during the summer months.

About 1300 the Ute Indians (who never numbered much more than ten thousand strong) arrived in today's Colorado Rocky Mountains, probably forced into their move by the stronger and more populous Plains Indians. The Utes loved what they called "the Shining Mountains," and it wasn't long before they roamed from present-day central New Mexico to Wyoming and from today's Front Range of Colorado into the central section of present-day Utah. The Utes generally traveled in small family units of five to ten people. They would winter with other Ute groups in the low valleys and then break up and travel into the higher areas in the summer. Because of the rough terrain, harsh weather, and lack of game they never really lived year round in the San Juan Mountains. They were usually just passing through. By the early seventeenth century the Utes had obtained the horse and many European implements, including guns and blankets, from the Spaniards, making their life immeasurably easier. However, they had a very uneasy truce with the Spanish, sometimes living at peace and at ease with the new culture and at other times waging bitter war.

Each Ute band had its own favorite territory, but they often traveled into the lands controlled by the other Ute groups. They were brave and ferocious fighters who didn't hesitate to defend their homeland from other tribes or from the Spaniards coming up from the south. Spain claimed the San Juans from the time they were first visited by Coronado in 1546, while he was looking for the legendary Cibola—the Seven Cities of Gold. The name of the mountains themselves comes from the Spanish word for St. John—probably a reference to the wilderness travels of John the Baptist. It wasn't until the late eighteenth century that Spanish exploration began in earnest and even then most of the explorers skirted around these high and rough peaks.

In 1765 Juan Maria de Rivera made one of the first recorded explorations through the San Juans. Rivera (and later the Escalante and Domiguez expedition) relied heavily on Ute guides. He probably came from Santa Fe over Cochetopa Pass and then followed the Gunnison River to near present-day Delta. It was a place where the Utes came together for their great councils—an easy place to find because it was where two large rivers (the Gunnison and the Uncompahgre) meet. It was also a good spot to spend the winter because there was little snow and it lay at a low elevation. At the rivers' confluence Rivera carved his initials, a cross, and the date on a young cottonwood tree, and he soon thereafter returned to Santa Fe with ore samples.

Most of the mountains and rivers had already been given Spanish names by the time of the Escalante-Dominguez expedition of 1776. It was proof that many Spaniards had already been in the mountains. However it wasn't until the 1820s that the name *San Juans* came into common usage. The two friars, Escalante and Dominguez, were looking for a better route to California when they passed near present-day Telluride, down the San Miguel River, over Dallas Divide, and down to near the present site of Delta. There they found the cottonwood tree with Rivera's initials. Don Bernard

Miera y Pecheco was the official cartographer of the expedition, and he drew a surprisingly detailed and accurate map of the San Juans, even though the expedition only skirted the mountains. Therefore, he probably was given detailed information by others who had traveled into the San Juan Mountains. A map published by Humboldt in 1811 and the map by Don Miera referred to the San Juans as "the Sierra de las Guillas" or "Mountains of the Cranes"—a reference to hundreds of thousands of cranes that passed through the eastern flanks of the San Juans in the San Luis Valley in the warmer months.

We have no record of most of the Spanish explorers in the San Juans because they were in the mountains illegally. But later a Spanish coin dated 1772 was recovered seven feet underground by a man digging a cellar in Howardsville. There are many signs of early Spanish prospectors, such as tunnels, arrastras, tools, and coins. Official expeditions had to give one-fifth of all that they found to the crown. No one wanted to give this money to a faraway king if they could keep from it, so these incursions were made in secret.

There were also French explorers in the southwestern San Juans in the 1790s. (The land was claimed off and on by both Spain and France during this period.) A group of 300 Frenchmen traveled from near present-day Leavenworth, Kansas, to determine the mineral worth of New France. They mined in several areas but found most of their gold near Wolf Creek Pass in the southeastern part of the San Juans. They might have also made occasional trips into the area around Baker's Park.

French, Spanish, and American fur trappers were in the San Juans as early as 1811 and began to move into the mountains in force in the 1820s after Mexico gained its independence from Spain. The fur trapper trade didn't last long (basically 1824 to 1845), but it did much to open the land to further settlement. Many of the trappers became guides for later expeditions. Perhaps the most famous trapper, Kit Carson, spent considerable time in the San Juans. An injury obtained when his horse fell in the San Juan Mountains led to Carson's death at an early age.

Antoine Robideaux was another famous San Juan trapper and trader. He was born into a French family in St. Louis but early in his life moved to Santa Fe. In 1828 he established a trading post called Fort Uncompahgre near the spot of Rivera's cross, near the confluence of the Gunnison and Uncompahgre Rivers. It served as a link with Sante Fe and a supply point for the fur trappers who had spread out over the high mountains looking for the elusive beaver. Beaver was in high fashion for making gentlemen's hats. Their pelts were waterproof and rugged and worth their weight in gold. Robideaux's fort proved to be very successful for many years.

Just like the Native American, the trappers traveled lightly over the land, so very little remains from their time. Eventually, beaver hats fell out of favor, and the bottom fell out of the pelt market. As prices fell the Utes felt that they were being cheated. They couldn't understand why they were receiving less for their pelts. They eventually rebelled and in 1844 burned Fort Uncompahgre to the ground. Then the floodwaters of the Gunnison River finished off the job, and absolutely nothing remains of the original Fort Uncompahgre today. There is, however, a wonderful replica of the fort. Ken Reyher has written an interesting book entitled *Antoine Robidoux and Fort Uncompahgre*, which goes into detail about this fascinating era in San Juan history.

In 1833 sixty fur traders led by William Walton of the St. Louis Fur Company were in the present-day Rico area and traveled as far north as Trout Lake-perhaps even farther. In 1839 another trapper, Thomas Farnham, wrote "among these heights (the San Juans) live the east and west bands of the Utahs. The valley in which they reside are said to be overlooked by mountains of shiny glaciers, and in every way resemble the valleys of Switzerland." There were many other trappers in the San Juans, including William Becknell and "Peg Leg" Smith.

In 1848 the United States won its war with Mexico, and the San Juans officially became part of the United States. However, as if to prove that the mountains wouldn't be conquered easily, John C. Fremont's disastrous 1848 expedition into the San Juans in the middle of the winter cost ten men their lives. The expedition failed even though the famous mountain man Old Bill Williams was their guide, filling in for Kit Carson who had a major illness at the time. The speculation is that Williams didn't turn soon enough to go over Spring Creek Pass and ended up in extremely rugged territory. Eventually the group detoured to the south where a man named Stewart found placer gold, chose to keep his find a secret, returned at a later date, but couldn't find the site again. Fremont later called the San Juans "one of the highest, most rugged, and impractical of all the Rocky Mountain ranges, inaccessible to trappers and hunters even in the summer time."

In 1853 Captain John Gunnison came across Cochetopa Pass, skirting the San Juans to the north in his search for a railroad route. He was convinced that the route was not favorable, and Gunnison eventually was killed by Piute Indians in Utah. However, it would prove to be the easiest of all the passes in the San Juans. Gunnison's second-in-command, Lt. E. J. Beckwith, later wrote "the Indian name for range west of the San Luis Valley is Sahwatch, but it was more generally known by the Spanish name of San Juan." In 1857 Randolph March and his group of about sixty men barely made it across Cochetopa Pass in December.

The Utes, at this time, were under the very obvious impression that the land they occupied was theirs. They had absolutely no fear of the whites, and they had total control over the San Juans. The U.S. government made the first of many Ute treaties in 1849. The treaty stated that the San Juans were in Ute territory, but the Utes would be peaceful toward American citizens and allow them to pass through their land. Such a "solemn" promise didn't last long because gold was discovered in 1858 on the eastern slopes of the Rockies. The next year a hundred thousand fortune seekers rushed into what would become Colorado—all hoping that the land would prove to be as rich as California did a decade earlier. Very little placer gold was found, and many of the men quickly returned to the East. However many prospectors turned to other portions of present-day Colorado, including the San Juans, to seek their riches.

The flood of inhabitants wasn't just Americans. A large number of Mexicans came to ranch and farm in the San Luis Valley and the eastern slopes of the San Juans. By the 1860s many of them had joined the Americans who came to explore for gold. Many other Mexicans took up support positions for the prospectors such as farmers, ranchers, and merchants. In fact the small settlement of La Loma was already established across the Rio Grande River when the Americans established their first San Juan settlement (Del Norte) in the San Luis Valley. Once gold was discovered it would become a major supply point for the San Juans. Mexicans served as railroad workers, household servants, and employees on the Ute reservations.

In 1859 a group of prospectors left the Pike's Peak area, worked their way down the Gunnison River, and then went up the Uncompahgre River where they found encouraging amounts of precious metals. The *Santa Fe Gazette* printed a San Juan prospector's journal in the spring of 1860. Albert Pfeiffer and Henry Mercure spent a few weeks in the San Juans and reported many gold deposits. A few of the disappointed prospectors on Colorado's eastern slope therefore seized on the hope that they might make a rich discovery in southwestern Colorado.

In July of 1860 Charles Baker and six other men came over present-day Cinnamon Pass, and made it into what is now called Baker's Park. The men spent the last part of July, August, and some of September placer mining. These six men included the future Baker's Park pioneers W. H. Cunningham and George Howard, as well as men called Bloomfield and Mason, and Samuel Lyor and J. Purdon. Cunningham spent most of his time in the gulch named after him, Mason prospected in Arrastra (originally called Mason) Gulch, and Baker spent his time in Eureka Gulch. Because they were placer mining, the men spent considerable time sawing lumber and creating sluices. They did no prospecting at all for mineral veins and they sent word back to their financial supporters that they were getting twenty-five cents in gold from every pan. This discovery was not a rich find of placer gold, but it was respectable.

Baker and his men went to Abiquiu, New Mexico, in September; came back to the park to establish a townsite called Baker City and a toll road during October; and then traveled over to the San Miguel River, down the Dolores River, and back to Abiquiu in November to spend the winter. At this time Baker saw a considerable number of Utes who were not pleased with his presence.

The *Rocky Mountain News* of October 12, 1860, carried an article about Baker's Park that declared that "the metalliferous development of (the San Juans), if not the North American continent, reaches it culminating point in this region." Such a statement caused quite an excitement back along the Front Range of Colorado, and from October through December of 1860, several parties of prospectors, totaling about 150 to 200 men, left Denver for the San Juan area. Most men traveled via Taos or Abiquiu New Mexico, and they built their road and bridges as they went along. Some of these men made it to the Animas Valley and wintered at what became the first Animas City at the extreme north end of the park (about twelve miles north of present-day Durango). Others went to Abiquiu or Taos to wait out the winter. Still others were on the road during the entire winter.

The *Rocky Mountain News* of November 9, 1860, warned its readers:

Go slow, very many of our people are becoming excited on the reported richness and promise of the San Juan mines. It is evident that there is "speculation and profit" in the enterprise for a few who are already interested in these discoveries, but nothing tangible and authentic has reached us of sufficient weight to warrant the rush to the mythical El Dorado.

Further warnings were printed in later editions but were pretty much ignored. Most groups waited until early spring, had a difficult journey, and discovered only small amounts of gold along the Animas River. Other men were probably inspired by and looking for the gold that Stewart had found with the Fremont party. A few men came into the San Juans via Canon City by going down the Gunnison and up the Uncompahgre to the area around present-day Ouray. Santa Fe and Canon City both touted themselves as the "Gateway to the San Juans." Although most parties carried only a few months' supplies, a few parties brought several wagons of supplies pulled by oxen.

A party of prospectors spent the winter of 1861-1862 in the bowl that now holds the City of Ouray. Some of these men might have come from the Baker party, but O. H. Harker, who came to Ouray in 1861, stated that his party came from Fort Garland by way of Cochetopa Pass. Harker was quoted as saying that "it took nerve and staying qualities to prospect in the San Juans in those days."

Baker wrote the *Santa Fe Gazette* in November 1860, making the highly exaggerated claim that he had made rich strikes and that "there will be not less than 25,000 Americans engaged in mining and agricultural pursuits...within a year, perhaps double that number." Baker further wrote that the only feasible way in was by Santa Fe and Abiquiu. Even more men came to southwestern Colorado, and by spring as many as a thousand men were posed to enter the mountains at the spring thaw, even though discouraging reports were still coming into Denver.

In early March 1861 Baker went back to the park named for him, and a few weeks later a large party of men and women were at "Camp Pleasant," which was close to the present-day Durango Mountain Resort. Due to the rough terrain, their wagons had to be abandoned at this point and many of the women and children stayed there while the men went ahead to explore. When prospectors started coming into the park that spring, they found Baker already settled into his brush shanties. He had filed placer claims at the mouth of Cunningham, Eureka, and Arrastra Gulches, but his results were not encouraging, averaging only about fifty cents a day (instead of the reported twenty-five cents a pan).

George Howard took a party from Camp Pleasant over to the Dolores River near present-day Rico. They split into two groups, some prospecting south down the Dolores and some going north to the San Miguel River. Meanwhile, the main base camp was moved from Camp Pleasant back downstream to Animas City, where many log cabins were built and even some farms started—though they were in the middle of Ute territory and the land was a total wilderness.

The Utes were only fairly tolerant of the whites, and then only when the whites "paid them" with supplies. Thomas Pollach, who had brought in large amounts of merchandise for the prospectors, traded for four Navajo children that the Utes had taken as slaves and made sure they all found good families to live with. Pollack also gave away a large portion of his supplies trying to keep the Utes peaceful, but the Utes kept warning the whites to get off their land. They often followed up on their warnings. It was estimated that as much as ten percent of the white population in the San Juans in 1861 was killed by Utes—most of them on the road between Animas City and Abiquiu.

All the early San Juan prospectors were so focused on finding gold that they ignored the very obvious signs of silver. There was a very practical reason for this oversight—silver in the San Juans is usually found in galena (a silver-lead ore). Silver was only worth one-twentieth an equivalent amount of gold; therefore, twenty times as much was needed to be of equal worth. A very valuable amount of gold ore could easily be transported by burros or mules, but a silver-lead ore would require that very large amounts of the heavy ore be shipped—almost an economic impossibility without good wagon roads or railroads nearby, neither of which existed. The San Juans were extremely remote, still located in Ute territory, and it would cost a small fortune to bring the silver ore out to market on the backs of mules and burros.

Besides those men on the Animas River, the Doc Arnold party, composed of twenty-four men coming from Denver, had ended up on the Uncompahgre River, where they spent the winter of 1860-61. In the spring they traveled south on the Spanish Trail down the San Miguel and Dolores Rivers and eventually ended up at Animas City. That spring, another party left Baker's Park, went through the Ouray area, and prospected around Cow Creek.

In the meanwhile most of the men in Baker's Park in the spring of 1861 became disgusted with their meager discoveries. Many left and went back to Denver over Stony Pass or Cunningham Pass (which were already well known as a shortcut out of Baker's Park), and others went back down the Animas River to Animas City. The Civil War was looming on the horizon, a San Juan provision train had been attacked by Indians, sickness set in, and the prospectors almost starved. Many of them dreaded another winter in the wilderness. Most of the prospectors abandoned Animas City. Twice, Baker's party threatened to kill their leader. A rumor spread that he had been hanged. It was generally considered that the entire "San Juan Excitement" was a scam, and by June of 1861 there was a steady exodus away from Baker's Park.

In June 1861 George Gregory wrote the *Rocky Mountain News* a letter that was published under the headline "The San Juan Humbug." In the letter Gregory described the extreme hardship of travel into the area and the hazards of the country. He said that men were recovering very little gold and that Baker was "almost a lunatic." He advised that no one should even think of going into the San Juans.

Nevertheless in 1861 another major rush had occurred—as many as one thousand men were reported to have been in Baker's Park that summer. They came from all directions—Denver, Del Norte, Canon City, Abiquiu, Taos, and Arizona. These hopeful men found mining prospects worse in Baker's Park than they had been on the eastern slope. There were simply too many men and too little gold—at least the placer gold, which was all the men were looking for. Baker stole away from the park during the night and soon thereafter became involved in the Civil War. It was not long before most of the men were involved with the American Civil War, not panning for gold in the San Juans. In December of 1862 a Captain Smith and six men ventured into "the heart of the San Juans"—presumably into the Baker's Park area, and they supposedly built a log cabin. They were having a little luck prospecting until Utes killed two of the party and forced the others to leave. It would be almost a decade before prospectors returned to the San Juans again in any numbers.

Baker came back for a short while in 1867 but was killed in Arizona by Indians later that year. The richest pans of gold after a day's work at Baker's Park had still only totaled $2.50. No one paid attention to the heavy black stuff in the pan, which was later determined to be rich amounts of silver.

In 1863 another treaty was made with the Utes (although only the Tabeguache band was actively represented). The Utes were moved off the eastern slope of Colorado but were promised that the part of Colorado west of the San Luis Valley would remain theirs forever. The United States was involved in the Civil War and basically was just trying to keep the Colorado Utes peaceful.

In 1864 Robert Darling and a group of Mexican army officers prospected near present-day Rico. They obtained permission from the Ute Indians to be in the area but found little and soon left. Ironically rich strikes were made at this very same spot some two decades later. Once again, the reason for the disappointment was that the party was looking for gold nuggets such as had been found in California.

The treaty of 1868 (or Hunt Treaty) set out the first clearly defined boundaries of Ute territory. The Utes basically agreed to stay in western Colorado, which the United States again solemnly agreed they could keep forever. The treaty would last only five years before the whites were back at the bargaining table.

Calvin Jackson left Prescott, Arizona, for the San Juans in 1869 with a party of twenty-two men, and Jackson's group combined with Captain Cooley's expedition of twenty-eight prospectors. After Indian troubles arose all but eight of the fifty men turned back to Arizona. Those who decided to keep going were Dempsey Reese, Adnah French, J. C. Dunn, N. Marsh, David Ring, Wood Dood, A. Lomis, and "Old Boston" Graves. By the time they reached the Dolores River, winter had already set in, so they decided to go to Abiquiu. They didn't know it, but there was another small party of prospectors camped for the winter just a short distance up the Dolores near present-day Rico. There were also a few other prospecting parties that wandered through or near the southern San Juans in the early winter of 1869.

The Reese party ended up in Santa Fe, gained the financial support of the governor of New Mexico and others, and then left for Abiquiu in the spring of 1870. There they met with a party of Utes, who agreed to let them prospect in their territory if they would not plow the ground, build cabins, or make fences. The original party had added four other men to their group and in April they split into two groups and headed back into the San Juans. Dempsey Reese, Miles T. Johnson, Abnah French, and Thomas Blair came to Baker's Park and made rich silver discoveries, such as the Little Giant and the Mountaineer Mines. The other party went up the Dolores but had little success. Both groups went back to Santa Fe for the winter, and the next spring they all returned to Baker's Park.

By late 1870 there was a total of about fifty prospectors in Baker's Park. A considerable amount of mining was done in Arrastra Gulch, so named because of a Spanish *arrastra* (or a crude crushing mill) erected there by the Reese party and used during 1871 and 1872. Most of the ore for the arrastra came from the Little Giant Mine. William Mulholand, James Cook, and Thomas Blair (all pioneers of Silverton) also obtained interests in the Little Giant. It was obviously the main mine of the area at the time, and it produced the long sought-after gold. Other claims were also filed, including the Pride of the West Mine in Cunningham Gulch and the Aspen Mine on Hazelton Mountain.

There were many ominous meetings, like this one, between the Utes and the white men that were not supposed to be on their land. From Harper's Weekly. Author's Collection.

In the winter of 1870-71 the discoverers took specimens from the Little Giant to Governor William Pile of New Mexico and convinced him to help finance the prospectors. In the meanwhile Special Indian Agent William Arny, while visiting the Utes in 1870, noted that prospectors had gone up the Animas to Baker's Park and also up the Dolores River to what became Rico. He warned the whites that Utes might allow prospecting but would object to any attempts at settlement. In his report he estimated two hundred prospectors to be in the San Juans and noted that 274 mining claims had been filed. He even included a map of the San Miguel Mining District on the Dolores River! Yet as winter approached, every one left the mountains again.

By the spring of 1871 many prospectors were looking for lode claims in the San Juans because the rich lode strike at the Little Giant was highly publicized. A party of twenty-five to fifty men came to Baker's Park as soon as they could travel the Cunningham Gulch route (which was becoming a popular approach) from the little settlement of Loma, which would soon become Del Norte.

Exaggerated reports were that the Little Giant ore was now running as high as one thousand dollars to four thousand dollars per ton of hand-sorted ore concentrates. Miles Johnson crushed twenty-seven tons of ore in his arrastra in 1871. The arrastra was a large wooden circle and run by water power. The entire structure was built with a hand ax, plane, and one-inch auger. The men at the Little Giant mined a quartz vein, ground the gold out, and then panned for gold. That year the men only milled about four thousand dollars in gold, and their costs were almost as high. However, it was a good start!

In August 1871 Chief Ouray and about twenty other Utes visited the Baker's Park mines. There was no trouble, but he complained about the presence of the whites to the Indian agent, who took steps to notify the miners not to trespass on Ute land. When winter came the prospectors again left the mountains, probably by Stony Pass or Cunningham Pass. Del Norte was now a well-established small town, and a few of the prospectors spent the winter there. Others went to Denver or even back East.

In spite of the Indian agent's warning, in the spring of 1872 Del Norte was filled with prospectors waiting for the snow to melt. Estimates of the number of men ran as high as two thousand, although the actual number was probably closer to five hundred. Some of the first men over Stony Pass built a small community in the vicinity of the Little Giant. There were no more than a dozen tents, and a few cabins and the prospectors did not name their little community.

At the same time George Howard (a member of the original Baker party) built a cabin at what would later become Howardsville. George Howard was an interesting man. He was a bachelor who lived with a number of cats, which looked as if they could have been wild at one time. He kept his cabin spotless and covered the floor with handsome animal pelts that he had trapped and tanned himself.

John P. Johnson, Charles Clase, Richard Corley, and R. J. McNutt built cabins and filed on a mill site that same year near what became Eureka. Dempsey Reese also built a cabin in Arrastra Gulch, where a large group of prospectors celebrated the Fourth of July.

Another 1872 group included George Ingersoll, Billy Quinn, John Dunn, Andy Richardson, and Edwin Wilkinson. The men of the group staked many claims—Quinn, Dunn and Richardson discovered the famous Highland Mary. In Arrastra Gulch they located the Sampson and built another arrastra. Still another group, Lindley Remine, George Lang, Bill Rump, and George Hamilton prospected near the junction of Howard's Fork and the San Miguel River near what later became Ophir. They found good amounts of placer gold. Remine came back in 1873 and built a cabin near Remine Creek (the first cabin in San Miguel County) and located the Navika Placer—the first mineral claim in the San Miguel District.

Exciting mineral discoveries were being made all over the San Juans and tents sprouted up all over the mountains. Many of the pioneer settlers in the San Juans arrived during this time. However, the Ute Indians still occupied the land, and the prospectors had to live in very temporary quarters. They were in a land where there were no roads and no way to replenish supplies without traveling for days. Furthermore, their own government was constantly trying to remove them from land that the U.S. government had promised the Utes would be theirs "forever."

By late 1872 the Little Giant's owners were bringing in the parts for a stamp mill. Major E. M. Hamilton brought them over Stony Pass by ox team. The men of the Little Giant transported a twelve horsepower engine and one thousand feet of cable all the way from Del Norte so they could have a tram from the mine to the new mill. Neither the mill nor the tram was set up that year as they arrived too late in the summer. The flood of prospectors increased so rapidly that it was necessary to send U.S. troops to try to maintain peace between the whites and the Ute Indians. In late 1872 a sawmill was brought to Baker's Park. This action really infuriated the Utes because it showed that the whites intended to stay. Something had to be done!

Ostensibly, the military was in the San Juans to help keep the whites out of Ute territory (a deadline of June 1 was even set for their removal). In reality they were there to protect the whites from the angry Utes. A total of about 150 men prospected all summer in 1872, and about 1,500 claims were filed. Most men left during the winter of 1872-73, but a few stayed all winter.

The government's focus on the San Juans shifted near the end of 1872, and efforts were soon under way to remove the Ute Indians rather than the prospectors. By 1873 it was time to make it clear that they had to give up their "Shining Mountains." As a result of the first two Ute treaties, Chief Ouray had been made the head of all the Ute tribes. Now he balked on giving up any further land. He asked the Americans if their chief was not big enough to live up to his promises. He reminded them that they had twice, by treaty, recognized that this land belonged to the Utes. The whites by their own law had recognized the validity of their title. The U.S. government was stymied about what to do.

By the summer of 1873 there were several thousand men prospecting in the San Juans. A few mining districts were even formed, and it was evident that the prospectors would prevail over the Ute Indians because it was generally known that there were several enormous and extremely rich silver veins in the area. The Little Giant's stamp mill wasn't in full operation until July 19, 1873, but they still shipped out twelve thousand dollars in ore in 1873. The little mill also processed a little ore for some of the other area mines. About two thousand in ore specimens were also taken that year-to show to prospective investors.

In the summer of 1873 the federal government sent a reconnaissance party to see just how many prospectors had settled in the San Juans and to determine exactly what the area had to offer. They were also prepared to move the prospectors out of the mountains if treaty efforts failed. The head of the party, Lt. E. H. Ruffner reported:

> The origin of the reconnaissance was the disturbed relations between the Ute Indians and the miners of the so-called San Juan district. This district was reported as embracing the claims located on the Animas River, and on the Lake Fork of the Gunnison River. These districts, formerly opened and abandoned, had become again the centers of wild speculation, and prospectors were reported as rushing there from all quarters. To the Ute Indians, occupying a consolidated reservation indefinitely large and embracing certainly one portion of the field and possibly all, the prospect of a wild flood of white men occupying their lands without regard to their guaranteed rights was anything but pleasant, and they early protested against the invasion. An attempt was made, in the summer of 1872, to secure a cession from them of the disputed territory. It was a failure, however, and when the rush of miners in the spring of 1873 promised to be greater than usual, the remonstrances of the Utes grew to threats during the winter, and they firmly said that the miners must leave or war would follow.

Finally, Felix Brunot, one of the commissioners appointed by the U.S. to negotiate with the Utes, came up with an idea. Years earlier Chief Ouray had his only living child stolen by the Arapahoes. Ouray had not seen the child since he was taken, but the boy was left-handed and had a scar on his chest, which would help the government agents to find him. Ouray agreed that if the government would find his son he would convince the Utes to give up their land. His son was found, and Ouray did as he had agreed. Ironically, Ouray's son refused to admit that a Ute could be his father, but there was little doubt of paternity. The man had the identifying scar on his chest, was left-handed, and bore a definite resemblance to Ouray. Was Ouray a traitor to his people? Perhaps, but the Utes felt these steep mountains to be nearly worthless, the white man's gold and silver meant nothing to them, and the whites already occupied much of the land in question.

As soon as it became obvious that the Utes would be moved out of the San Juans, the government lifted its order banning prospectors, and they began to flood into Baker's Park. More than two thousand claims were staked by the end of 1873, and the Little Giant was sending out good amounts of valuable ore. The discovery of mines that later became world famous were made in the San Juans that year-the Sunnyside, Shenandoah, Silver Lake, and Gold King to name just a few. Albert Burrows and others prospected all the way to the Mineral Point area in 1873. Five or six men stayed and worked the Little Giant during the winter of 1873-74, and several other men also worked other prospects that winter. The treaty with the Utes was made in September of 1873, but it had to be ratified by the U.S. Senate before it became official. That wouldn't happen for six more months, but nevertheless a small rush was on for Baker's Park.

The San Juan Triangle was a name given in the early mining days to a highly mineralized 250 square mile area formed by a roughly southwest to northeast line running through what would later be the towns of Telluride, Ouray, and Lake City, then extending down on both sides to Silverton. This area proved to be so mineral rich that most of the five thousand square miles in the San Juans that were formed into its fifty mining districts fell within its boundaries.

The logical way into the San Juan Triangle was from the west, but the Utes were still in control of that territory, so the prospectors had to come from the east over rugged Stony Pass or from the north through Lake City and over Engineer or Cinnamon Pass. It was about 125 miles away, but Del Norte was the nearest supply point, bank, newspaper, and post office to Silverton. Saguache was the supply point for Lake City at the northeast end of the triangle.

Although the 1874 Hayden Survey declared Cinnamon Pass too high and rugged for a wagon road, Otto Mears evidently didn't get the message. He followed right behind the expedition building a rough road. Mears's road became an alternate route into Silverton from Saguache, but the upper section really wasn't much of a wagon road. It was completed in 1877 and served along with Stony Pass as a rough, barely passable wagon road into Baker's Park.

In April of 1874 the U.S. Senate ratified the Brunot Treaty, and that summer the San Juan Mountains were legally opened to prospectors. The thirty-ton Greene Smelter was erected on Cement Creek near Silverton that year, but it didn't process ore efficiently and its expenses exceeded its income. The mill was later dismantled and moved to Durango. About five hundred men prospected in Baker's Park in 1874. Ironically, the Little Giant, the mine that started the San Juan rush, closed in 1874 in part as a result of legal problems with its owners but also because its vein was fast disappearing. On a positive note, the men working on the new toll road to Silverton from Saguache discovered gold near Lake City.

Although discoveries of great potential were being made, men didn't just flood into the San Juans. The terrain was simply too rugged, and the winters were too severe. Travel in the high mountains was possible only three or four months during the summer. It was only during this short time that the rich mineral deposits high up in the mountains could be found. New discoveries couldn't be made in ten feet of snow!

Even with all its rich ore, the mineral production of the San Juans began very slowly because it took capital and a transportation system to ship the ore. In all the San Juans no ore was officially shipped in 1874, then only $90,517 worth in 1875, $244,663 in 1876 (the year Colorado became a state), $377,472 in 1877, and $434,089 in 1878. A great deal of ore was being produced, but very little of it was being shipped because of the high cost of transportation. The San Juans were isolated by distance, terrain, and climate. As one disgruntled prospector told the *Denver Tribune:* "The San Juan is the best and worst mining country I've ever struck. It has more and better mineral ... but you can't get at it ... and when you're in, you see, you're corralled by the mountains, so you can't get your ore out." A burro could carry only 150 pounds and a mule 250 pounds of ore. The long, rugged trip to the nearest mills could take weeks and cost five dollars per animal, thereby making it unprofitable to ship anything but the very richest ore. The rest of the ore just piled up on the dumps outside the mine, where some of it still sits today. It was a problem that was to plague the San Juans right up into current times.

As men poured into the new mining districts many hastily drawn guidebooks were printed to help the newly arriving prospectors. *William's Tourist Guide and Map of the San Juans* reported "the surface croppings are so distinct that, without previous experience, the would-be prospector will invariably discern the 'signs' and with pick and hammer secure a piece of 'blossom rock' which assures that he has 'struck it rich.'"

It wasn't anywhere near that easy, but the guidebooks served their purpose, which was to secure a steady stream of greenhorn prospectors along the railroads to the new mines. Frank Fossett summed up the problem:

> Promising as were the numerous discoveries of the San Juan country in 1873-74-75, they were generally of no immediate benefit to their owners, on account of the distance from an ore market, wagon roads, and railways. The region labored under peculiar disadvantages. It was made up of almost inaccessible mountain ranges, and at that time, was so remote that capitalists and mill men were not inclined to investigate its mineral wealth. The pioneers who had been making discoveries of rich veins were too poor to build works for extraction of the precious metals, and it cost too much to get ore to market to admit of attempting it, unless it was wonderfully rich and money was at hand to defray shipping expenses.

However the prospectors did persist, the capitalists and the railroads eventually arrived, and the miners eventually took huge sums of ore out of the mountains. In the last century, the fifty mining districts that existed in the San Juan region produced almost a billion dollars in gold, silver, lead, zinc, copper, and other precious metals.

These minerals would be worth many billions more at today's metal prices. Geologists generally believe that large quantities of ore still exist in the San Juans. They just haven't been discovered yet, because they don't rise to the surface. In some cases minerals are known to remain in existing mines.

Most of the hardrock San Juan mines are now closed because of low metal prices, the high price of production, and the long distances to mills and smelters. Serious ecological concerns also impede the resumption of mining. However, some day the price of precious metals will rise high enough to again spark mineral production in the San Juans. In the meanwhile the mountains play host to hundreds of thousands of tourists that arrive each summer to visit some of the most beautiful places on earth—most with a history that is almost as rich as the minerals that have been found here and those that still remain.

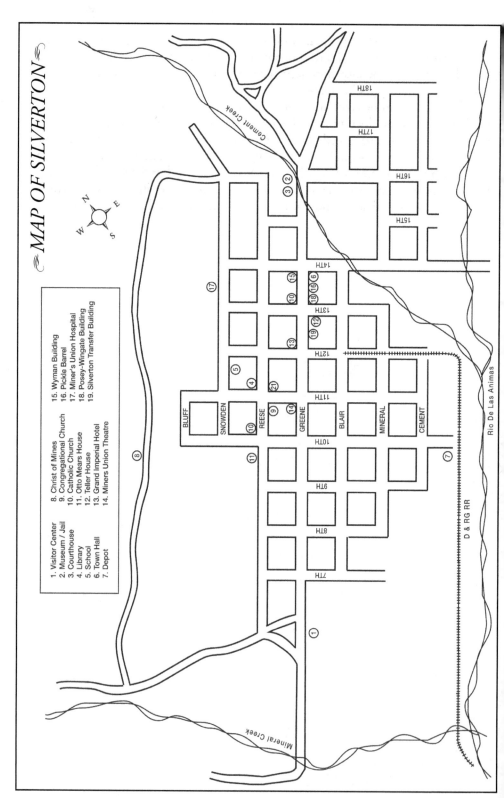

MAP OF SILVERTON

1. Visitor Center
2. Museum / Jail
3. Courthouse
4. Library
5. School
6. Town Hall
7. Depot
8. Christ of Mines
9. Congregational Church
10. Catholic Church
11. Otto Mears House
12. Teller House
13. Grand Imperial Hotel
14. Miners Union Theatre
15. Wyman Building
16. Pickle Barrel
17. Miner's Union Hospital
18. Posey-Wingate Building
19. Silverton Transfer Building

Cement Creek

Mineral Creek

Rio De Las Animas

D & RG RR

BLUFF
SNOWDEN
REESE
GREENE
BLAIR
MINERAL
CEMENT

7TH
8TH
9TH
10TH
11TH
12TH
13TH
14TH
15TH
16TH
17TH
18TH

⌐ *Chapter Two* ⌐

THE TOWN OF SILVERTON

Baker's Park (which encompasses about two thousand acres and is shaped like an hourglass) is located in the central San Juans and at the southern tip of the San Juan Triangle. Glacial scouring formed the park, and then the valley filled with gravel as the glacier melted. To the northeast lies Boulder Mountain and behind it Storm Peak (13,487 feet), to the southwest Sultan Mountain (13,368 feet) and Grand Turk (13,087 feet), to the northwest Anvil Mountain, and to the southeast Kendall Mountain (13,066 feet). From the broad and flat park, prospectors spread up Mineral Creek toward the Red Mountains, up Cement Creek to Poughkeepsie, and up the Animas River to what would become Mineral Point. Because most of the early prospecting in the San Juans was done in this high (9,330 foot) mountain valley, it was logical that the first real town, Silverton, was eventually built at the location. It is the oldest continuous settlement of any size (the current population is about 500) in the San Juans.

The identity of owner of the first house in Silverton is a matter of great confusion. In 1872 Francis Snowden built a rough cabin in the area that eventually became the town site of Silverton about 1872. However, he had no interest in laying out a town site or having his cabin located in a town. Snowden was looking for his riches in the ground—not from selling ground. Tom Blair built a cabin in 1872 near the point where Cement Creek comes into Baker's Park, but his cabin did not end up within the city limits. Blair did consider founding a town but never followed through with the idea. Blair also built a cabin on the site now occupied by the Grand Imperial Hotel, but whether it was finished before Snowden's cabin is unclear. N. E. Slaymaker also built a house about this time, but it is generally agreed that it was the second house in town. A 1905 Silverton newspaper article stated that the first house in Silverton, as well as the first house in Baker's Park, belonged to John P. Johnson and that it was built in 1871, but there is no other proof of this fact.

In October 1873 Dempsey Reese, Tom Blair, and William Kearns claimed as homesteads all the land that is now occupied by Silverton. Their purpose was not to farm or ranch but to secure the land for a town site. Kerns and Blair never followed through, but Reese filed on an eighty-acre parcel. He drew up plans for the town and even built a fence around his house to enclose his livestock, but his home was not within the present-day city limits. Silverton was originally laid out on paper in 1873 using the name of "Quito." The names "Reeseville" and "Greenville" were also used. Late in the year Reese induced a Mr. Tower, who was accompanied by his nineteen-year old bride (who was probably the first woman in Baker's Park), to build a sawmill in the townsite. Reese also induced George Greene to open up a smelter in the new settlement. It was a wise move because these two events secured the future of Silverton.

The Colorado territorial legislature felt the San Juan area to be growing so fast that it needed its own county, so on February 10, 1874, La Plata County was carved out of Lake and Conejos Counties. The small settlement of Howardsville was named the county seat, because the town of Silverton did not exist at the time. The original La Plata County was quite a large area and contained most of the "San Juan Triangle."

In June of 1874, Dempsey Reese, F. M Snowden, Thomas Blair, N. E. Slaymaker, W. J. Mulholland, K. Benson, and Samuel Champlin formed a town site company. Later the names of George Greene and Co. and Calder, Rouse and Co. were added. Both were smelting firms that had opened establishments in or near the new town in response to Reese's offer of free land. An official survey for the townsite was not done until September 9, 1874.

The origin of the name *Silverton* is not certain, but a prospector commenting on "we may not have gold, but we have silver by the ton" is a good guess. The name certainly points out that the local citizens had finally recognized it was silver, not gold, that would be the basis of the area's riches.

After the signing of the Brunot treaty in April 1874, settlers and prospectors flooded into Baker's Park. Some estimated as many as two thousand people visited the area that summer. Although Howardsville had been designated the county seat of La Plata County in February, the locals wanted Silverton to be the county seat because the sawmill and two new smelters (the Greene and the Holmes) had been built there. An election was held in the summer of 1874 to decide the location of the county seat. There were five potential sites on the ballot, but Silverton received 183 votes out of a total of 341 cast.

The Greene Smelter's initial run in 1874 didn't work out well, but it was remodeled and did better in 1875. Two other smelters were soon built nearby. None of the smelters worked well, but the Greene was the best.

There were several problems with the Silverton smelters. They needed lots of silver-lead flux to work correctly and there just wasn't enough of that ore available. The cost of importing coal to Silverton was high enough that they couldn't smelt at a profit. It also seemed hard to get a smelter to work properly at high altitude because of the lack of oxygen.

However, the Greene sawmill, which was attached to the smelter, helped produce much of the lumber used in Silverton during the first few months of its existence. The town grew fast—so fast that rents were extremely high. The town's four north-south streets were named after founding fathers—Greene, Blair, Reese and Snowden. The east-west streets were numbered First through Eighteenth (although First through Sixth no longer exist.)

By the end of 1874 George Greene had his cabin, store, and nearby smelter. L. L. Ufford also occupied a cabin, which also served as the county clerk's office, district court, and sometimes as a post office. The Silverton cabins were just scattered around the townsite. Some of the prospectors who left the area for the winter gave Ufford forty gallons of whiskey with instructions to sell it all by spring. He had no problem doing so, although some accounts say Ufford bought all the whiskey himself and gave it away! Rhoda, of the Hayden Survey, wrote that in the winter of 1874-75 there were about a dozen houses available for nearly fifty men and eight women who stayed in the isolated, high mountain valley. Here in the wilderness, dances were held and debate societies met, although drinking water was delivered to town by dogsled. Pete Schneider brought the water in barrels from a spring at the base of Kendall Mountain and sold water for fifty cents a bucket. Because there were so few buildings, some houses were occupied by as many as three families.

This early drawing of Howardsville and Silverton condensed the distance between the two towns. Frank Leslie's Illustrated Newspaper, May 8, 1875. Author's Collection.

Mail was starting to be delivered to the town, but it was sporadic and expensive. At first individual contracts were given to anyone willing and able to make the trip to Del Norte. Usually they were allowed to charge whatever the traffic would bear. Twenty-five to fifty cents a letter seemed to be the going rate—at a time when most postage was only a penny or two. In the winter this meant the mail had to come in over Stony Pass on snowshoes, so it was delivered on only a couple of occasions. There was no actual post office building so mail was just carried around until it was distributed to the proper recipients.

By spring of 1875 the town's provisions became very scarce, and many citizens came close to starving before relief supplies finally arrived. By summer the new settlement really began to swell. Large numbers of businessmen moved to the new town. The *La Plata Miner* printed its first edition in Silverton on July 10, 1875. John R. Curry was the editor. It later combined with the town's second newspaper, the *Silverton Standard*, and is now the oldest newspaper in continuous operation west of the Continental Divide in Colorado. By the time the paper opened there were twenty houses in Silverton as well as a large assortment of tents. That same month the Rough and Ready Smelter opened within the city limits. There was not yet an actual business district. Stores and saloons were scattered all over town. A school building (which doubled as a town hall, church, and meeting hall) was built during the summer of 1875, and Isaac Grant opened the first hotel that summer. There was still no post office, although an official postmaster had been appointed.

A big Fourth of July celebration in the summer of 1875 culminated in the town's first shootings. A drunken miner by the name of Kelley accidentally shot John Conner. Kelley was chained to the floor of a cabin (since there was no jail) until his trial (in those days trials were held within a few days). The shooting was ruled an accident, and Kelley was released. During the celebration that ensued, another shooting occurred; the charges against the defendant later had the charges dropped without a trial.

The first white child was born in Silverton in 1875; Frank Harwood was to live in Silverton for all of his long life. At one time he carried fifty to one hundred pound packs of mail and supplies on snowshoes from Grassy Hill, on the other side of Stony Pass, to Silverton. Construction also proceeded in 1875 on the road between Silverton, Howardsville, Mineral Point, and over Cinnamon Pass to Lake City in an attempt to make it passable for wagons.

On January 31, 1876, San Juan County was carved out of La Plata County and, at the same time, Silverton was officially named the county seat, even though it had already been serving in that capacity for some time. Mail now came in regularly from Lake City over Cinnamon Pass. There were also efforts being made to make travel easier to Durango. A road was completed from that city to near present-day Durango Mountain Resort but from that point on there was only the old Ute trail that went down Bear Creek. The May 6, 1876, issue of the *La Plata Miner* showed just how bad the isolation was.

> Last Tuesday afternoon our little community was thrown into a state of intense excitement by the arrival of the first train of jacks, as they came in sight about a mile above town. Somebody gave a shout, 'turn out, the jacks are coming, and sure enough there were the patient homely little fellows filing down the trail. Cheer after cheer was given, gladness prevailed all around, and the national flag was run up at the post office. It was a glad sight, after six long, weary months of imprisonment, to see the harbingers of better days, to see these messengers of trade and business, showing that once more the road was open to the outside world.

The Fourth of July 1876 was celebrated in a big way that year. Not only was the town proud of its accomplishments, but Colorado became a state, and it was the centennial celebration for the United States. The celebration went on for several days, as did most of the Fourth of July celebrations for the foreseeable future. Celebrations were easy to come by any time of the year as Silverton now had ten saloons. Prostitutes also showed up in town for the first time. By the end of summer 1876 the Town of Silverton boasted three hundred to four hundred people, about one hundred houses, two sawmills, and four stores.

By fall of 1876 the Greene Smelter was up and running again using its new technique. Water jackets were used for a lining instead of firebrick; the latter lasted only a few firings, and the bricks were expensive. Only galena ores were being worked with any success, and the smelter produced only about 140 tons of bullion running about five hundred dollars per ton. Shipping costs remained high-an average of sixty per ton in 1876 and fifty-six dollars per ton in 1877. Very little had been done to the "roads" into town other than removing the downfall, stumps, and the largest rocks. It was almost impossible for a wagon to make the journey without a breakdown.

In the winter of 1876-77 the snow got so deep that the mail was often brought in by dogsleds by a man named Snyder. He supposedly rewarded his big, black dogs with a drink of whiskey after each trip. Mail carriers William Moore and Charley Bates complained that the men of Silverton were advertising back East for women companions and that there was "five tons of pink, green, and blue envelopes piled up at Aldens (a stop on the eastern side of Stony Pass) and the place smells of musk and patchouli like a ten-cent barber shop."

Estimates of the number of people that stayed in Silverton during that winter of 1876-77 range all the way from 150 to 350. Everyone expected pack trains by March or early April, but it was late May before they made it through the deep snows, and Silverton's population was again very close to starvation.

In the spring of 1877 enough new entrepreneurs came to Silverton that a central business district started to form on Greene Street between Thirteenth and Fourteenth Streets. A fire brigade was started to try to bring some semblance of safety from fires. Efforts were made to start a church. There had been services the previous year, but no preacher was present. Rev. George Darley fought his way over Cinnamon Pass in a snowstorm by following the mail carrier but his efforts to establish a church weren't successful. Other ministers, including Parson Hogue from Ouray, came on occasion but no church was built nor were any regular services held.

There were nearly three hundred buildings in Silverton by the end of 1877, the Denver & Rio Grande Railroad (D&RG) had surveyors in the area looking for a route into town for the railroad, and the toll road from Durango was being used; but most businessmen and many of the residents still left town for the winter. The D&RG investigated routes via Cunningham and Stony in 1876 and 1877 but came back later to survey the route up the Animas in 1879. The winter of 1877-78 was another snowy and cold one, and by December there were already many shortages.

Although Silverton wasn't as bad as the wild and wooly frontier towns portrayed in the movies, it did have its fair share of violence and hardships in its early days. Nor were local gunfights as romantic as the movies would make them out to be. In October of 1878, Tom Milligan and Bill Connors were involved in Silverton's first gunfight. The men had been feuding, met on Greene Street, and drew on each other. Milligan hit Connors in the stomach with the first shot and then followed the crawling Connors and fired at him twice more (both shots missed). Connors died three days after the shooting. Milligan was immediately arrested and brought to trial the day after the shooting. He was found innocent by reason of self-defense. His acquittal occurred two days before Connors died! "Justice" was obviously a lot swifter in those days, and sometimes the locals refused to wait for "official" justice. There was an active vigilante group in Silverton that tried to help keep the peace. "Cap" Stanley, who opened the first brickyard in Silverton, was supposedly the leader.

In 1878 the toll road was officially open along the Animas River to Durango. It started near the Champion Mine south of Silverton and then crossed the Animas River many times until it quickly rose two thousand feet in elevation and came out near Lake Electra. The toll was three dollars. That same year Otto Mears also completed the toll road from Saguache to Lake City to Silverton. However the preferred route into town was over Stony Pass to Del Norte. There were now three ways—although all very rough—into Silverton, but there was virtually no travel in to or out of the town in winter except for a very hard and dangerous trip on snowshoes or skis.

Although freight costs dropped, the cost of processing ore wasn't any better at the local smelters. The average smelter cost was one hundred dollars per ton and only a small percentage of the valuable minerals was recovered. The mine owners using the local smelters still had to ship the refined metals out before they could sell their product. The biggest problem was the lack of development capital. The San Juans were rich with ore, but no one had heard of the new mining district and it was going to take money—and lots of it—to get the silver and gold out of the hard quartz veins.

The Greene Smelter shut down and moved to Durango in late 1878. It had produced about 870 tons worth $376,000 in the five years it had been in existence. The move to Durango was due to many factors—an abundance of nearby coal, better results at lower altitudes, and cheaper labor, to name a few. Silverton was left without a smelter that could refine ore at an acceptable level.

In the summer of 1879 Bill Harwood and a crew of about forty upgraded the trail over Stony Pass. It was still a hard road to travel, but it was a good enough wagon road to cut freight rates from Del Norte to thirty dollars per ton. This amount was half the original cost but still high enough that it allowed only the richest of ores to be shipped at a profit. Harwood also built a road toward Ophir but made it only to Burro Bridge before stopping for the winter. He finished the Ophir Pass Road in 1880.

In 1879 violence again raised its ugly head in Silverton. On the night of May 27 James Dermody bit night watchman Hiram Ward on the finger and pushed him into a ditch. His brother, Pete Dermody, also joined in the fight. The night watchman drew his gun and killed James, but he was found innocent at his trial. (It seemed awful hard to get a conviction in those days!) Then on August 23 James M. Brown, one of the owners of Brown and Cort's Saloon, was killed. Brown had tried to get one of his drunken customers, Harry Cleary, to leave. Cleary shot Brown in the chest, then Brown took ten wild shots at Cleary as he was running away. One of Brown's wild shots hit Hiram Ward, who shot back and put another bullet in Brown. Cleary was arrested and jailed but before he could be tried, he was taken by a vigilante group and hung from the ox-shoeing frame at the back of the blacksmith's shop. (It didn't seem too hard to get hung in those days!)

Fossett in 1879 had optimistic words to say about Silverton:

This is a growing and prosperous mining town.... In the lofty mountains that overhang the park are numberless mineral veins, some of great size and many extremely rich in silver.... Snow falls to a tremendous depth during the winter months, and avalanches occasionally sweep travelers on this trail down mountainsides.... Population 1,000.

In 1879 Silverton started experiencing problems with arsonists. The Reese Hook and Ladder Company had been formed in 1878, a wagon and other fire fighting equipment were ordered from Denver, and the volunteer fire department formed. There was quite a celebration when the fire equipment arrived over Stony Pass. "Half the male population of Silverton" went to the top of the pass to help bring the hook and ladder wagon down the steep slopes. The local firefighters were all volunteers and not only fought fires but participated in firefighting races and contests all over the state. Nevertheless, quite a few Silverton buildings were to burn down over the next few years.

This photo is looking northeast down Greene Street in the 1880s. The Grand Imperial Hotel is on the left. Author's Collection.

It wasn't until 1879 that regular mail came into Silverton during the winter. One man would take the mail from Silverton to the Highland Mary Mine, and then another man would take it over the top of Cunningham Pass to what was called Grassy Hills, where it joined the regular route down to Del Norte. The mailmen, traveling on snowshoes, would also carry supplies to the miners who were holed up along their route. Snyder and his dogsled team (which usually carried water) were pressed into service to carry the caskets for winter burials. The editor of the *La Plata Miner* reported that, in his opinion, the dogs were better than men in carrying mail and supplies.

On April 10, 1881, Charley Moorman got drunk and at 2:00 a.m. randomly fired into the Coliseum Gambling and Dance Hall. He hit J. K. Prindle who was watching a faro game. Moorman was captured, taken by a mob estimated at from fifty to three hundred men, and hung from a tree. On August 24, 1881, yet another shooting occurred. Kid Thomas, Bert Wilkinson, and Dyson Eskridge arrived in town from Durango. They were followed to Silverton by Luke Hunter, the La Plata County Sheriff. He enlisted the help of Marshall David Ogsbury and the two went to the Diamond Saloon to arrest the men. The gang started firing, killing Marshall Ogsbury. Kid Thomas turned himself in, but the other two escaped. Once again the citizens of Silverton didn't wait for "official" justice. A vigilante committee was formed, and the captured man was taken from the jail by a lynch mob and hanged. The two men that escaped headed toward Castle Rock near Durango. Bert Wilkinson was turned in for the twenty-five hundred dollars reward money that had been posted. Once again vigilantes stormed the jail and hanged the alleged killer. The coroner's jury the next day supposedly formed the verdict that the man "came to his death from hanging around." The third man was shot to death just a few days later in Durango. Silverton vigilantes had now hung three men in less than five months.

It wasn't until the early part of the 1880s that Silverton truly began to prosper. In 1880 the Thatcher Brothers Bank opened a bank (the Bank of Silverton), and the First National Bank (originally opened as the San Juan National Bank) opened in 1883. The banks gave the town a sense of permanence. The first stagecoach arrived in May 1881

This photo looks southwest, diagonally across Greene Street about 1883. Three different delivery wagons are visible. Author's Collection.

(although wagons had made it much earlier). In 1880 telephone service was started between Silverton, Ouray, and Lake City. There were also connections at Capitol City, Rose's Cabin, Mineral Point and Animas Forks. The Martha Rose Smelter opened in 1881, but it operated for only a few days before shutting down. No one knew it at the time, but it was not to reopen until ten years later, when it was called the Walsh Smelter. But most of all prosperity came because of the arrival of the D&RG Railroad on July 13, 1882. The railroad followed much of the old toll road route into Silverton, thereby eliminating wagon road access from Durango for many years. Excitement grew throughout the San Juans. The *San Juan Herald* proclaimed Silverton to be the "Gem City of the Mountains, the most prosperous and promising camp in the San Juans." The effect of the D&RG's arrival at Silverton cannot be stressed too much. Freight costs for ore and concentrates fell from thirty dollars to twelve dollars per ton. The cost of shipping freight to the San Juans had a corresponding drop in cost. Until 1882 only the richest ore could be shipped. Now even low grade ores could be shipped.

All the activity prompted the *Denver Tribune* on September 21, 1882, to note that Silverton had 350 buildings and many business houses. "Perhaps no town in Colorado boasts of a more reliable and enterprising class of businessmen, merchants and banks."

Perhaps because of all the newly found prosperity and respectability, George Brower, who owned the Arlington Saloon, hired Wyatt Earp to run his gambling hall during 1883. Wyatt had a bad reputation (this was two years after the famous gunfight at the O.K. Corral), and there were even rumors that he was wanted for murder at the time. After Earp had been in town for a while, another rumor had it that he had joined with Bat Masterson, Doc Holliday, and others to return to Dodge City (which had banned them), but the May 12, 1883, *Silverton Democrat* reported:

© Colorado Historical Society

This little Silverton girl stood very still for the photographer about 1885. Anvil Mountain is in the distance. Courtesy Colorado Historical Society. Photo by Miller and Chase. (X4656)

The Club Rooms of the Arlington are conducted by Wyatt Earp who is a pleasant and affable gentleman, and not mixed up in the Dodge City broils, as the Denver papers would have us believe, and the editorial paragraph in the *Denver Republican* of the 16th, placing him at the scene of action and counseling the citizens of Dodge to take the law in their own hands, is wholly unwarranted, from the fact that Mr. Earp is now, and has for the past three months, been a peaceable and law-abiding citizen of Silverton, and not been in Dodge City for the last four years.

However, just a few days later Bat Masterson and Doc Holliday did arrive in Silverton, seeking Wyatt's help, and all three left for Dodge City (where their problems were eventually settled peacefully).

An interesting description of the two faces of Silverton appeared in the 1883 *Durango Southwest:*

As night darkens, the street scene changes from the work and traffic of the day and assumes quite a festive tone. Sweetly thrilling peals of music are borne upon the night air, and the brilliantly lighted, palacial saloons are thronged by the sportive element, with the pleasure seeking and curious, all classes mingling happily. The sharper with his trap-game laid for the sucker just fresh from the hills with too much dust or bullion certificates, or the greeny from the east with more of Pop's bond coupons than he has of Ma's wit; either may be enticed into the glowing allurement.

Before the arrival of the train, business in Silverton fell off drastically in the winter. Most miners couldn't make it to town, supplies couldn't be brought in, and ore couldn't be sent out. After the arrival of the railroad many of the mines near Silverton built trams down to the rail line. Yet the railroad didn't solve all of Silverton's winter problems. During the winter of 1884 a heavy storm sent slides down all along the D&RG's tracks, closing the railroad for seventy-seven days (February 5 through April 22). The snow was so deep that it was impossible to bring in supplies by any means. Once again the town came close to starvation before the route reopened.

On June 28, 1885, another shooting occurred. Marshall Tom Sewall shot and killed a burglar at Kinnan's Hardware Store. The only problem was "he shot first and asked questions later." Many citizens felt he had violated the law himself by shooting without any warning. Then another murder occurred when L. F. Toles and John Barnet fought over a gambling debt. Although the two were best friends, Barnet stabbed Toles in the heart. Barnet was found guilty but was sentenced to only two years and nine months in jail!

Yet there was a better side to life in Silverton. A skating rink was flooded in early winter starting in 1885, and it continued for several years. There were parties for the women, and a reading library established for young men at the Silverton Club. The Silver Cornet Band played for special occasions. Silverton's town fathers made a weak attempt to clean up its notorious Blair Street in 1887. Saloon fees were raised to three hundred dollars a year, and women of easy virtue were not to "parade the streets, visit the saloons, or ride on horseback on any of the streets west of Blair."

As usual, Crofutt became a little carried away by the potential of Silverton in 1885:

> Its location is most desirable, on a high level above the river, with broad streets, bordered by shade trees and running water on each side; surrounded by grand old mountains carved in places with forests of timber, and filled with precious minerals of all kinds; a climate the most healthful; game in great variety for the taking; and scores of mining dependencies on all sides make its prospects for the future most promising.

The first true census was taken in Silverton in 1885 and showed a population of 1,195 (Crofutt had estimated 1,500). Although Silverton's economy was strongly based on mining, tourism was becoming another viable industry. Before the end of the century, Silverton was printing guidebooks, and scores of tourists were being brought in on the D&RG Railroad.

In 1890 electric service was provided to Silverton, but it was available only during evening hours. The original thirty-five member Cow Boy Band was moved to Silverton from Dodge City in 1890. The band had been formed in Dodge City in 1881. The trademark of the band was several sets of five to seven foot longhorn horns. The band played for the inauguration of President Benjamin Harrison in their ten-gallon hats, and they also played at the inauguration of Colorado governors Routt and Adams. The reason given for the move was the band would be closer to where the "real" cowboys were located, yet there probably wasn't a single cowboy in all of San Juan County. Otto Mears presented the band with a large banner (with his name prominently displayed) to carry in parades. In February of 1893 Buffalo Bill purchased the band and made it part of his Wild West Show.

About 1890 gold started turning up in the local millruns. It was a critical time for Silverton because the Silver Panic of 1893 would all but wipe out silver mining. Ironically gold, which had not been located in Baker's Park in the 1860s, was now

The men at the Hercules Mine Number 2 Tunnel had a great view of Silverton. The racetrack and baseball field are visible and there was plenty of room for the town to grow. From a postcard in Author's Collection.

The Cow Boy Band was brand new to Silverton as one banner promotes Otto Mears and the other refers to them as the Dodge City Cow Boy Band. Courtesy Colorado Historical Society. (X4738)

being found just as silver failed. A short rush even surfaced in 1893 when gold was found at Bear Creek about twelve miles east of Silverton. The "Bear Creek gold rush" quickly waned, but mining remained profitable around Silverton at a time when mines were folding all over Colorado. In fact some of the largest mines ever, such as the Silver Lake, were established at this time; and they were to remain in operation for decades.

The *Durango Wage Earner* wrote in April 1896:

> Silverton has more new buildings in the course of erection than any other town in the San Juans. She has more gold and silver lying around in the sides of her mountains than the world could coin into money in two or three generations. There are no flies on Silverton.

In 1899 the *Silverton Standard* published a booklet on Silverton that suggested the town's name might be changed to "Goldton." It went on to state that "San Juan County's outlook for the season of 1900 is decidedly bright; new mines are being opened up and new reduction plants in contemplation are to be erected."

By 1900 Silverton had become one of the largest cities in Colorado with three churches, one hundred businesses, and good schools. The local mines employed more than one thousand men year round and almost twice that many worked in the summer. Silverton was now served by no fewer than four railroads. The D&RG provided access to the outside world, the Silverton Railroad ran to Red Mountain, the Silverton, Gladstone and Northern ran to Gladstone, and the Silverton Northern ran to Eureka (it was extended to Animas Forks in 1904.) Silverton stood to gain from the production of all the mines located along these different railroads, and the wealth from over one hundred square miles of rich mining country flowed into Silverton over the rails of the little narrow gauge railroads.

In 1901 Silverton purchased the water works system, which had been owned by a private company before that time. Water was taken out of Boulder Creek and held in a reservoir north of town. The privately owned, coal-fired electric company (electricity was becoming common) and the sewer system were purchased by the town in 1902, making it one of the first towns in Colorado to own all three of these utilities. The town constructed a new power plant because the old one was unreliable. Silverton was a town that was truly booming. From 1900 to 1918 only Telluride or Creede could come close to rivaling Silverton in gold production. New banks and businesses were coming to town, and the population swelled to over four thousand. It was still a wide-open town, however. The saloons and bawdy houses prospered along with the local mines.

But not every business did well. When the Bank of Silverton (located in the Grand Hotel) was forced to close on January 2, 1902, its president, J. H. Robins, was found dead alongside a water tank on the D&RG. He had shot himself in the head. The *Silverton Standard* lost two thousand dollars when the bank closed. The competing *San Juan Prospector* consoled the Standard by asking, "What was a newspaperman doing with $2,000 anyhow? He ought to lose it."

Silverton, like the other San Juan towns, went through labor problems in 1902-04. The unions were strong in Silverton (over one thousand miners were members), but

This photo of Silverton was taken about 1900 from near the cemetery and shows the city's water reservoir. Courtesy Denver Public Library. Photo by L.C. McClure.(06387)

Quality Hill at the base of Anvil Mountain was the place to live in Silverton in 1900. There were even wooden sidewalks in front of some of the houses. Courtesy Colorado Historical Society. Photo by E. Adams.(x4401)

This photo of the Silverton depot gives some idea of how high the snow could pile up in Silverton in the winter. Courtesy Colorado Historical Society. (F44094)

there was no violence like in Telluride. As elsewhere, the mine owners eventually won out, and by 1910 the Silverton union had closed its doors.

On September 3, 1904, the Hub Saloon was held up a little before one in the morning. Herman Strobble, the manager, tried to stop the robber and was shot twice, but he eventually recovered. The robber fled out the back of the saloon. One of the robber's stray shots killed John Lotus, a miner, and another wounded the bartender, James Bothwell. The sheriff and a posse went after the man, and the next morning discovered his body. He had shot himself in the mouth with his .44 caliber Colt pistol.

The first automobile (driven by Dr. D. C. Mechlin) came to Silverton over Stony Pass on August 26, 1910. It had to be pulled the last few miles up the pass by a team of horses. Two women and a man waving an American flag met Mechlin at the top of the pass. They were just the start of well-wishers that lined the route into Silverton. Later several cases of dynamite were set off, and the city hall bell rang continuously for hours. The car was a Croxton Keeton-patterned after a French Flich by Renault. The next day the car went on to Ouray over Red Mountain.

In 1913 efforts were started in earnest to get a highway built from Silverton to Durango, yet it wasn't until 1920 that the first car made it over the new road. The Million Dollar Highway was dedicated in 1924, finally making it possible to drive all the way from Durango to Ouray.

In 1914 Colorado voted to ban the production and sale of alcohol, and on January 1, 1916, prohibition became the law. Colorado's prohibition came some four years before the Eighteenth Amendment made liquor illegal throughout the United States. It may seem strange that Colorado welcomed prohibition, but Silverton had voted to stay wet. Prohibition didn't work, and it became the era of speakeasies and bootleggers. Isolated mining towns such as Silverton became bootlegger havens. Duane Smith, in his *Silverton: A Quick History*, quoted one local, "Some of [the bootlegged liquor] would turn up your toenails, I'll tell you that much. Never killed anybody that I know of. Out of here though, I've heard it's blinded some people."

The local law enforcement didn't really try to arrest bootleggers, but federal agents would occasionally raid Silverton. Because there was no road from Silverton to Durango, the feds could come in only on the railroad, and as the agents left Durango someone would almost always call Silverton, and the saloons and speakeasies were back to serving soft drinks by the time the federal agents arrived in town. It was December 5, 1933, before Prohibition was repealed.

The 1918 "Spanish" flu epidemic struck hard in all the San Juan towns. The flu first hit Silverton on October 18 when two cases were reported. The town officials immediately closed all public places including churches, schools, and lodges; and all residents were requested to stay home. When they did go out, citizens wore white surgical masks. Yet despite the precautions, forty-two people died in a one-week period. The town hall was turned into a hospital. By November 2 the count of dead was up to 146 people.

During this horrifying period 1,298 Silverton citizens came down with the flu or related pneumonia. The *Silverton Weekly Miner* reported, "In all its history the San Juan has never experienced such a siege of illness and death." One strange but terrifying fact was that young adults seemed to be the most prone to die. Silverton gained the dubious distinction of having the highest percentage of its population die from the Spanish flu of any city in the United States. The national average was one-percent-still a major disaster-but more than ten percent of Silverton's popula-tion died within only three weeks. Gravediggers had to dig trenches at the cemetery instead of individual graves to keep up with the dead. The bodies were dug up later and reintered in individual graves. One trench contained sixty-two persons. The paper had no room for obituaries; it just posted alphabetical lists of the dead. Even the undertaker died of the flu.

Mining started declining in Silverton in the 1920s. A Silverton economic brochure from the 1920s pleaded, "We need capital for mining, milling, and cyaniding-the harvest is here, it is yours to reap, the earliest to come will have first pick; take head, and come first." By the 1930s, with the Great Depression in full swing, times were really bad.

The last long snow blockade of Silverton occurred in February 1932 and lasted some ninety days. Supplies and freight were brought in from Needleton by dogsleds and by a string of pack mules from Ouray, which made it into Silverton on March 4, 1932. This snowstorm also brought about one of the strangest mailings in U.S. Postal history. The *Denver Post* of April 7, 1932, reported:

> For the first time in the history of the post office department, it is believed, a ton
> of hay has been shipped by parcel post. This was revealed Wednesday in a report
> to the State Utilities Commission. The hay was badly needed to feed dairy cattle

in the snowbound town of Silverton and was ordered by the Mullin Lumber Company of Silverton from the Farmer's Supply Company of Durango. The Durango firm was unable to ship the hay by freight because the railroad stops sixteen miles from Needleton Pass. It was decided to throw the responsibility on Uncle Sam for the delivery of hay. The hay was pressed into bundles to conform with the maximum size and weight specified for parcel post and offered to the postmaster at Durango. The postage amounted to $14.00. When the hay had reached the end of the railroad it was transferred to Silverton by pack mule at a cost to the post office department of 5 cents a pound. Thus, the post office department lost $86.00 on the transaction but the cows in Silverton will munch hay for a few days at least, and the children of the isolated town will have fresh milk again.

Mining didn't do well during the Great Depression, but one man, Charles Chase, managed to keep the nearby Shenandoah-Dives Mine operating during the terrible economic times. He not only kept about 250 men working but also fed anyone who was hungry at his boardinghouse. His mine became a model of innovations and economies that many other mines were later to adopt. He even asked the Silverton merchants and landlords to lower their prices to help out. At a time when both prices and demand for ore were low, Charles Chase kept Silverton alive.

Mining rebounded sharply during World War II when demand for base metals soared, but at the same time it was a period of rationing throughout the United States. There was a new wave of prospecting throughout the local mountains and many abandoned mines were reexamined. Mining did okay for six or seven years after the war, but by 1953 it began another rapid decline. The Shenandoah-Dives closed that year. Silverton's population fell from 1,375 in 1950 to only 890 in 1957.

However, a different type of economic activity took up some of the slack—tourism became vitally important. Several movies were filmed in or near Silverton in the 1940s and 1950s, including "Ticket to Tomahawk," "Across the Wide Missouri," "Naked Spur," and "Denver and Rio Grande." In the 1950s the D&RG narrow gauge train also began to bring in larger numbers of tourists from Durango each year, promoting Silverton as a "Trip to Yesteryear." Mining rebounded again in the early 1960s when Standard Uranium (later Standard Metals) took over the Sunnyside Mine. It closed, however, in 1985. By that time the train and tourism were firmly established. Silverton also started promoting summer events such as Hard Rock Days, the Iron Horse Bicycle Race, the Kendall Mountain Run, and brass band concerts. Lately, Silverton is also attracting winter business with its new extreme ski area, Silverton Mountain, and the Colorado Outward Bound School. Upon this base Silverton continues to build up a core of year round residents.

Because there has never been a major fire in the town Silverton still has many of its original historic buildings, so a tour of the city is certainly worthwhile. There are so many buildings of historic note that we can only hit the highlights. An appropriate place to start is the Christ of the Mines monument, which overlooks the town from Anvil Mountain. To get to the shrine follow the road from Fifteenth Street. The last part of the trip is a short hike. The monument is a statue of Jesus, twelve feet high, hand carved in Italy from Carrara marble. Stones for the background and base of the shrine came from the old Fischer's Silverton Brewery. The niche is twenty-four feet high and forty feet wide along its base. The Catholic Men's Club erected the shrine in 1959 when it looked like mining (and hence the town) were dying. The statue cost five thousand dollars and weighs twenty tons! Father Joseph McGuiness planted trees behind the shrine where foresters told him no one could grow Scotch pines, but these trees grew! Just a short time later metal prices rebounded, and the American Tunnel was started in the Sunnyside Mine. Mining prospered for almost another quarter of a century.

Silverton's Congregational Church at 1060 Reese Street was established in 1876 and formally organized in early 1878. The Rev. Harlan P. Roberts was the first resident Silverton minister and served until 1882. The cornerstone of the church building was laid August 20, 1880, and a concert was held in the building on October 2, 1880; but the church was not officially dedicated until 1881. The bell was bought by the Sunday School children, and the steeple added in 1892. The adjoining parsonage was built in 1884. The Catholics and Episcopalians also used the church for many years.

Father Thomas Hayes celebrated the first Catholic mass in Silverton in 1877 (it was also the only Catholic service that year). Missionaries from Del Norte came several times each summer over the next few years. After the train came to Silverton, the priests came more often. The original wooden St. Patrick's Church was built in the summer of 1883, but the congregation was formed the year before. The new brick Catholic church at 1005 Reese Street was built between 1903 and 1905 (as much of the work was done by volunteer labor), but it was not dedicated until April 27, 1907. The brick church rectory next door was built in 1906.

The Fisher Building (Teller House) was built at 1250 Greene Street in 1896. Fisher also owned the Silverton Brewery located on the banks of Mineral Creek. The brewery was a large stone structure built next to Fisher's house-also large and of stone construction. The upstairs hotel was named the Teller House after Senator Henry Teller, who was always very supportive of mining.

The Silverton Courthouse at Fourteenth and Greene has fortunately retained its original splendor. It was started on August 10, 1906, and was completed in December 1907 at a cost of one hundred thousand dollars including furnishings. The gold-colored dome and clock tower, tile floors, marble pillars, and large oval rotunda are beautiful reminders of bygone days. When it was built, the Silverton Courthouse was considered one of the most beautiful in the nation. On the courthouse lawn is a monument studded with ore samples from more than one hundred local mines. The courthouse has recently been remodeled and is still used as the main county building. The restored clock tower bell tolls every hour on the hour.

The large three-story city jail located beside the courthouse was built in 1902 and has served as a museum and the home of the San Juan County Historical Society since 1966. It contains a good variety of exhibits on Silverton and the San Juans. San Juan County originally had a small log jail, then a small stone jail, and finally the large brick jail by the courthouse. It is well worth a visit! The first floor originally held an office, kitchen, sheriff's residence and a cell for women on the ground floor. An office, cells and a bath were on the second floor. In the 1930s it was used as a home for the indigent and then sat vacant until taken over by the historical society.

The old Caladonia boarding house was moved down from Minnie Gulch and is being reconstructed behind the museum.

Construction began on the "Thompson Block" in 1882 by the Thompson brothers, and it was officially opened April 19, 1883. W. S. Thompson was the royal perfumer to Queen Victoria. With the arrival of the D&RG, the Thompsons were convinced that Silverton would become the largest city in Western Colorado (it was for many years). At first the building at 1219 Greene Street wasn't meant to contain a hotel. The third floor was to be apartments and offices, yet the Grand Hotel (run by Summa and Roe) soon opened on the third floor and the building became known by that name for almost half a century. In 1909 the hotel was named the Imperial and then in the early 1950s the Grand Imperial. The Thompson brothers owned the building until 1922.

Originally, there were shops on the first floor (a hardware store, clothing store, mining supply, and general merchandise) and offices on the second, but almost immediately the county ended up with most of the second floor offices, and the building also became the courthouse. It is a massive four-story granite structure plus a basement. Originally, it was seventy-five by one hundred feet. In the 1890s the building was the center of Silverton social life, beautifully furnished with guilded mirrors, ornate chandeliers, and Victorian furniture. The hotel has been restored on the inside several times over the years, but the exterior looks almost exactly as it did when built. Originally, it had fifty-four rooms and three baths, but today it has about forty rooms, each with its own bath. John F. Kennedy and Marilyn Monroe were among the distinguished guests at the hotel.

Jack Slattery's Hub Saloon operated for many years twenty-four hours a day, seven day a week. The great bar in the saloon was made in Denver and shipped to Silverton in three sections on the railroad. The large glass mirrors behind the bar were made in France. It is reported that the origin of the song "There'll Be a Hot Time in the Old Town Tonight" began when a train porter at the hotel was discussing what would happen to him if his wife found him at the bar. In 1903 someone accidentally locked the safe in the Hub Saloon, and no one knew the combination or how to open it! A locksmith had to be brought in to do the job. Jack Slattery was a big supporter of Silverton's baseball teams and was also a partner with Otto Mears in several mining ventures. Slattery became a director in the Silverton and Silverton Northern Railroads and was elected a state senator.

The Miner's Union Hall at the corner of Eleventh and Greene Streets (1069 Greene Street) was constructed during 1900-01. The thirteen hundred union members met upstairs, and the Prosser furniture store and an undertaking business occupied the downstairs for many years. More than thirty thousand dollars was spent on the structure. Later, the building was purchased by the American Legion. It was used for many years for dances, benefits, and meetings. The upstairs is now used as a theatre.

The two-story red sandstone Wyman Building at 1371 Greene Street was built in 1902 for thirty thousand dollars. Wyman opened a restaurant in 1881, which was just a few buildings south of his new building. He was successful in the packing and freighting business and dedicated his new building to his burros and mules. He personally cut the

The Silverton City Hall was built in 1908. Although badly damaged by fire in 1992, it was restored and still stands. Courtesy Colorado Historical Society. *(F26375)*

bas-relief of the burro into the stone on the upper corner of the building. The building has been recently restored as a hotel.

The Silverton depot at the southeast end of town (Tenth and Animas Streets) was built by the D&RG in one week in July 1882. It was meant to be temporary, but it was never replaced until the railroad stopped using it in 1966. It had room for baggage and freight at one end, offices and supply rooms in the middle, and ticket sales and a waiting room at the other. It was rebuilt in 1976 and is now owned and operated by the Durango and Silverton Railroad.

Starting in 1877 and continuing for years afterwards, Silverton's Blair Street supported scores of prostitutes and saloons. The area was open twenty-four hours a day and lined with over forty cribs, saloons, dance halls, opium dens, and gambling houses that, at one time, were reported to offer as many as 150 girls. Most of the establishments were between Eleventh and Thirteenth Streets. Many shootings and hangings occurred and many fortunes were won and lost on Blair Street. Several big-name gamblers and prostitutes of the Old West worked the establishments along the street. The well-known houses and saloons were the Mikado, National Hall, Sage Hen, Bon Ton, and Zanoni-Padroni. Some of the better-known girls used colorful aliases such as Nigger Lola, Big Mollie, Diamond Kate, Orgeon Short Line, Tar Baby, and Diamond Tooth Lil. For many years the license fees paid by the "sporting houses" and saloons of Silverton contributed a major portion of the revenue collected by the city. The usual fee was about five hundred dollars a year for "sporting houses" and a five-dollar a month "fine" for each girl. One sign of Silverton's prosperity that the town fathers didn't boast about was that the city collected sixteen thousand dollars in saloon license fees alone in 1907. About half of the original Blair Street buildings still stand. Silverton also had true "sporting houses." Both boxing and horse racing were very popular, and some of the top fighters and race horses in the United States came to Silverton.

The first stone building in Silverton is now occupied by the Pickle Barrel Restaurant at 1304 Greene Street. It was built in 1880 by Fred Sherwin and W. F. Knowlton and lavishly outfitted with ninety feet of black walnut counters and French walnut trimmings. Several other stores occupied the premises, and it became the Iron Mountain Saloon in 1910. Prohibition ostensibly turned the saloon into an ice cream parlor.

The original Silverton City Hall was built in 1883 on Blair Street, but as Blair Street eventually became the red light district, no respectable woman would go near it. The new City Hall, at 1360 Greene, was erected in 1908-09 from dark red sandstone quarried from South Mineral Creek. The cost of construction was $14,500. During construction the four columns on the front literally fell off! A circular rotunda with gilded highlights and skylights rises for two floors through the interior. It was restored in 1976 and then badly damaged by fire in 1992 but has been beautifully reconstructed to its original appearance.

The Silverton Public Library was built at 1111 Reese Street in 1905-06 from a generous donation from Andrew Carnegie. It was dedicated July 25, 1906. The city council immediately responded by appropriating $1,500 for books and passing an ordinance establishing fines for overdue books. The library looks almost the same, inside and out, as the day it was first built.

The Miner's Union Hospital was completed June 7, 1909. It was built at 1315 Snowden on the site of Colonel F. M. Snowden's original 1874 cabin. Silverton had several smaller hospitals by this time, but the Miner's Union was the first of any size. Its first floor had a ward, six private rooms, and doctors' offices. Another ward, more private rooms, and an operating room were upstairs. The building was used as a hospital for thirty years, and then it went to the county, which still owns it but leases it out for offices.

The Posey and Wingate Building at the corner of Greene and Thirteenth Streets (1269 Greene) is probably the oldest brick building used for commercial purposes still being used on the Western Slope. It was built in 1880 across from where the city hall was later built. Oliver P. Posey and John W. Wingate used part of the building for their hardware store, and the First National Bank of Silverton occupied the corner from 1883 to 1934. The building took two hundred thousand bricks, and the front is solid cast iron (twelve thousand pounds that came over Stony Pass). The original upstairs apartments were quickly converted to accommodate the Metropolitan Billiard Parlor. Later the downstairs was used as a recreation hall. The little frame building next door is Silverton's oldest commercial building, built in 1876 and originally occupied as a general store.

The Silverton Transfer Building at 1142 Greene Street was billed as the fanciest livery stable in Colorado when Clinton A. Bowman built it in 1897. Bowman left Silverton to help found the National Biscuit Company (Nabisco). The building has been used as a livery for most of its existence and includes an elevator to the second level.

Knut Benson, who was using the profits from his Butterfly-Terrible Mine and Mill near Ophir, built the Benson Block at 1200 Greene in 1901. The downstairs was originally a saloon, and the upstairs contained several hotels over the years. Later the downstairs was used as an automobile garage.

The church at 1105 Snowden was originally built to be an Episcopal Church but was rented as overflow classroom space to the Silverton School. It opened as a church in 1901 and was the site of the funeral services for Otto Mears—Pathfinder of the San Juans, although Mears was Jewish. It has been used as a church by various denominations since that time. The belfry was originally part of the schoolhouse at Eureka.

These are but a few of the historic structures that still exist in Silverton. Many of the local guidebooks or magazines have a detailed discussion of the homes and business buildings that are still standing.

☞MAP OF LAKE CITY☜

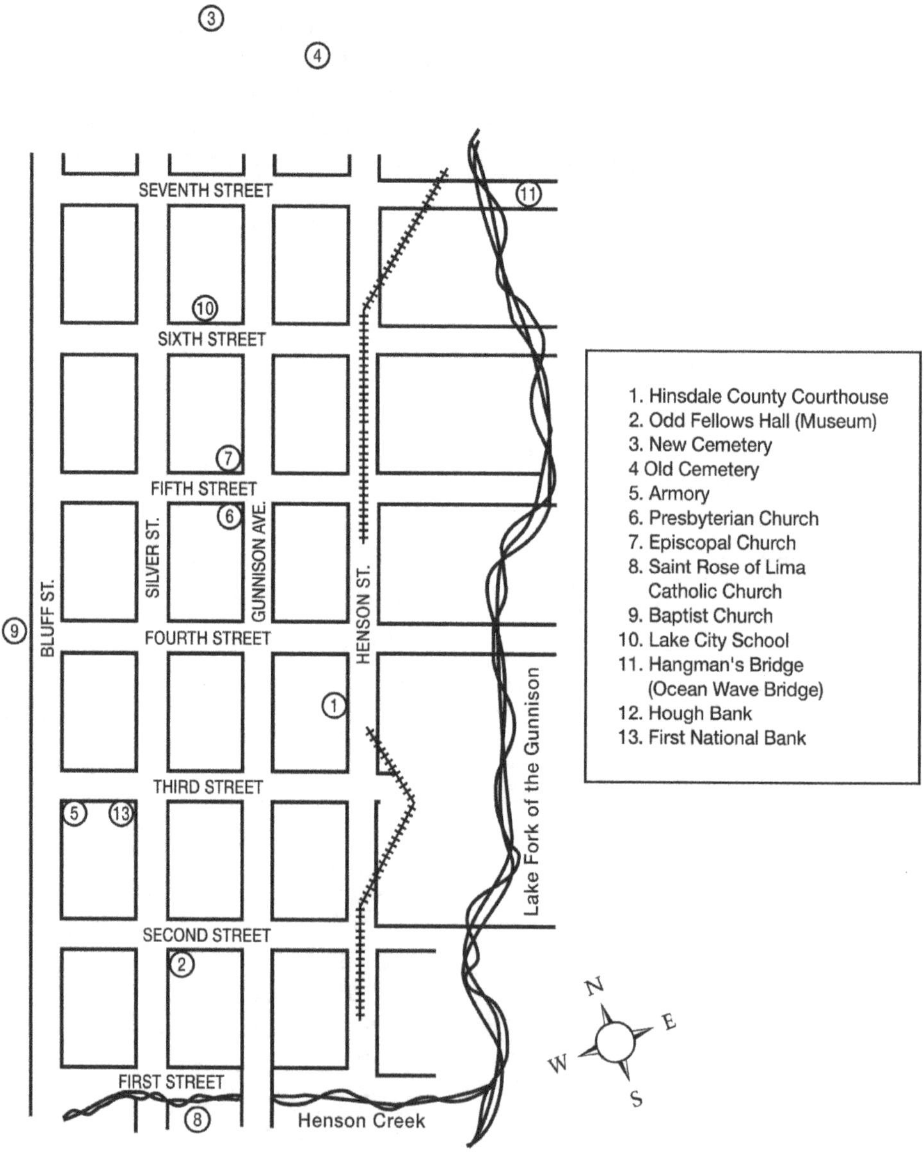

③

④

SEVENTH STREET

⑪

⑩

SIXTH STREET

⑦

FIFTH STREET

⑥

FOURTH STREET

SILVER ST.

GUNNISON AVE.

HENSON ST.

BLUFF ST.

⑨

①

THIRD STREET

⑤ ⑬

SECOND STREET

②

FIRST STREET

⑧   Henson Creek

Lake Fork of the Gunnison

1. Hinsdale County Courthouse
2. Odd Fellows Hall (Museum)
3. New Cemetery
4 Old Cemetery
5. Armory
6. Presbyterian Church
7. Episcopal Church
8. Saint Rose of Lima
   Catholic Church
9. Baptist Church
10. Lake City School
11. Hangman's Bridge
    (Ocean Wave Bridge)
12. Hough Bank
13. First National Bank

N
W   E
S

Chapter Three
LAKE CITY

It is very likely that in 1860 or 1861 members of the Baker party visited the site that later became Lake City, but no written records confirm that fact. We do know for sure that Colorado prospectors were in the mountains around present-day Lake City (which lies at an elevation 8,671 feet) by the summer of 1869, when James Harrison, J. K. Mullen, and George Boughton established a temporary camp at the mouth of Henson Creek. On August 27, 1871, Harry Henson (who had been one of the original Baker party) and three others discovered the Ute-Ulay vein. It was located about five miles upstream on the creek then named Godman (for Charles Godman, who was another of the other original Ute-Ulay discoverers). The creek was later renamed for Henson. The 1871 group had come over the mountains from Baker's Park, but they couldn't officially locate their rich claim or establish a settlement because they were prospecting on land that was still within the Ute Indian reservation.

Prospectors were back in present-day Lake City again in 1872. This group included Peter Robinson, O. A. Mester, and B. A. Tafts, who filed several mining claims. When Lt. E. H. Ruffner and his reconnaissance party came over Cinnamon Pass from Baker's Park during the summer of 1873, he reported at length on Granite Falls and then went up Godman's Creek. By looking at tracks he determined that no more than five or six prospectors had been up the creek; yet he also noted that it looked like the Utes had used the area extensively in the past (they had stripped the bark off the trees for food and medicinal purposes). However, none of the signs of Ute occupation looked very recent.

The Utes gave up the San Juans in 1873, but there was no great rush to the northern part of the San Juan Triangle. However because of the Utes' removal, Hinsdale County was formed on February 10, 1874, from parts of Costilla, Conejos, and Lake Counties. The county was named for George A. Hinsdale, a prominent Pueblo attorney and former lieutenant governor of Colorado, who had died the previous month. The county originally included present-day Mineral County. The first county seat was not Lake City but rather a place called San Juan City, which was a small settlement over the Continental Divide in Antelope Park, which was on an upper branch of the Rio Grande River. The small settlement was located about twenty miles above present-day Creede.

Lake City came about because of a toll road and Otto Mears. The toll road was built because of a rivalry between Del Norte and Saguache for the title of "Gateway to the San Juans." Almost from the discovery of the first San Juan ore, Del Norte had shipped supplies to Silverton by burro and pack trains via Stony and Cunningham Passes. Saguache counted on the efforts of its local merchant, Otto Mears, to make sure that its businessmen would also ship supplies by wagon to Silverton. Mears had already constructed a part of the route as a toll road to take supplies the original Los Pinos Indian Agency, which was located near the top of Cochetopa Pass. Mears hired California veteran road builder and miner Enos Hotchkiss to finish his "Saguache and

San Juan Toll Road." The toll road's route would eventually cover 130 miles, and the first 100 miles were useable by early August of 1874. It was the easiest and safest route into the northern San Juans, and one of the main reasons that the town of Lake City would grow quickly after its founding.

When in the Lake San Cristobal area, Hotchkiss succumbed to his prospecting instincts and took time off to look for mineral deposits. When he came across some promising mineral veins, he and his friends Henry Finley and D. P. Church filed the Hotchkiss claim. Gold ore from this mine later assayed at forty thousand dollars a ton - it was a very rich strike! However, Hotchkiss didn't know just how rich it was at the time of the discovery. After filing his claim, Hotchkiss came four miles down the Lake Fork of the Gunnison River to the large flat area that would become Lake City. Here Hotchkiss built two crude log cabins (presumably to claim the land as a town site) and then hurried to catch up with the rest of the road builders who were already some distance upstream, heading toward Silverton over Cinnamon Pass.

Hotchkiss, by the way, was about as tough as they came. He and his brothers loved to hunt mountain lions—one year killing thirty-two of the beautiful beasts, including one he lassoed and killed with a knife. Hotchkiss also had a knack for picking good townsites. A family named Bartloff, the Lee brothers, B. A. Sherman, and Finley Sporling were among the first to build cabins at the spot where Hotckhiss had contructed his cabins. Because the original discoverers had failed to follow up, Joe Mullin, Al Meade, and C. E. Goodwin also relocated the Ute-Ulay Mine during 1874.

News of Hotchkiss's discovery spread quickly, and a small flood of prospectors headed toward what would become Lake City. A reduction works was even built in 1874. Enough people were living in the settlement that on February 23, 1875, a special election was held and the county seat was moved to Lake City from San Juan City (known today as Dabney's). Lake City wasn't very big but San Juan City had only one merchant (who also acted as a postmaster), two cabins, a tent, and the log cabin courthouse. It was difficult to have a county that spread out over both sides of the Continental Divide, so the eastern part of the county was later split off and made into Mineral County after rich gold strikes were made at Creede in 1892.

In the early spring of 1875, the settlement of Lake City consisted of only thirteen log cabins and adobe structures plus a few tents, but during that year the town grew quickly. A town site company was set up in early 1875. Henry Finley was named president, and Hotchkiss and Mears were on the board of trustees. One of their first acts was to remove squatters from the land upon which the town site company had just filed, but Hotchkiss and other original road builders received a number of lots, presumably to compensate them for any prior claim that they had against the land. Women arrived early in the town's history, and 1875 saw the birth of the county's first white child and the first wedding. The first of Barlow and Sanderson's stagecoaches from Saguache also arrived over the new toll road on July 11, 1875. The stage continued to make the trip three times a week in good weather. After the discovery of rich ore and the establishment of Lake City, Del Norte responded by starting construction of its own road to Lake City, which tied in with the Saguache road on the other side of Slumgullion Pass. When the Lake City and Antelope Park Toll Road was completed November 22, 1875, Lake City was also connected to Del Norte.

Lake City had the first large "official" Fourth of July celebration on the Western Slope in 1875. There was no American flag to be found any where in the town, so one was made out of red flannel underwear, a blue flannel shirt and some white handkerchiefs. The festivities included a parade, contests, fireworks, and a public dance that lasted all night. (Lake City also celebrated Christmas 1875 in a grand manner. Three hundred eleven people attended a community supper on Christmas evening. Gifts were given to all the children present, and candy, nuts, and fruit were presented to everyone.)

By June 19, 1875, the Lake City *Silver World* newspaper was in operation with Harry Woods and Clark Peyton as editors. Otto Mears helped start the paper, because he knew it would help lure men to the region. At first the paper was located in a log cabin, although by May 1876 it was relocated to its own building between Silver and Gunnison Streets on Third Street. As with the newspapers in most mining towns, the *Silver World* spent a great deal of time praising the mildness of the local climate and the richness of its surrounding mines. The *Silver World* also reported the theft of horses, burglaries, and several gunfights during 1875. It reported only three stores, a restaurant, and a saloon in town at the time it arrived. The "houses" were all log cabins with dirt floors, mud roofs, and only two had any glass in the windows. There were no physicians in early day Lake City, but the paper reported that the city fathers felt the doctors weren't necessary because it was such a healthy place to live! The *Silver World* was the first newspaper on the Western Slope of Colorado and is still operating today.

On June 26, 1875, the Hayden Survey arrived in town. On July 1, 1875, the first post office opened. With the arrival of a sawmill later in the summer of that year, the first frame buildings began to replace the crude log structures. Lake City was officially incorporated on August 16, 1875. By November 1875 Lake City contained sixty-seven finished buildings and a population of about four hundred.

Lake City grew quickly in 1876. Since the Ute-Ulay mine sold for $135,000 in 1876 and the Ocean Wave started producing well, it was established that there were rich minerals in the nearby mountains. Overnight, Lake City ceased to be just a stop along the toll road to Silverton, came into its own right, and experienced a real building boom. By November of that year all of the original lots had been sold, and Lake City boasted almost one hundred businesses, seventy-six finished frame buildings, and forty-five more structures in the process of being built; the foundations for another seventy-nine buildings had been laid. So many people were moving to Lake City and lots had become so scarce that lot prices soared. The town now claimed eight hundred to one thousand year-round inhabitants, and more were arriving daily. The Lake City *Silver World* declared, "Homes are going up so rapidly that a citizen leaving town for a day's exercise has to inquire of his own domicile in the evening."

The Hinsdale House, Lake City's largest hotel at the time, was built in the summer of 1876. Other Lake City hotels included the American House built in 1876, the Occidental (all three of its locations were destroyed by fire), and the Pueblo House. By July houses were being built along the Lake Fork of the Gunnison and up Henson Creek. Two new banks were also built. The Hinsdale Bank was the first bank on the Western Slope. Merchants streamed into town, bringing every type of luxury item with them. Almost all of Lake City's visitors noted that the town was very refined for a mining camp that was only two years old. There were jewelry

53

stores, fine restaurants, and enough women that there was a ladies millinery store. Many professional men and their families lived in the town. It took four sawmills, a planing mill, and a shingle mill to keep up with all the new building going on, and telegraph service, which was installed in 1876 by Western Union, helped everyone keep in touch with the outside world.

One man wrote that Lake City "men are intelligent, even aristocratic, many of them quote Shakespeare." The Reverend Darley in his book *Pioneering in the San Juans* wrote that "among the men who came to the camp were many who had been trained in fine Eastern homes."

The Hotchkiss (now called the Golden Fleece), Ute-Ulay, Ocean Wave, Golden Wonder, Black Crooke, and many other mining properties were expanded. All these claims were within a few miles of the town. The Crooke Smelting Works and Van Gieson Lixiviation Works were open and doing a landslide business. The Crooke brothers were probably more responsible for Lake City's growth (and perhaps that of the San Juans) than any other factor. Even with rich minerals and easy transportation, capital was needed to work the new Lake City mines, and the Crookes supplied it. They owned the Little Annie and Golden Queen Mines when they opened their smelter, and later they bought the Ute-Ulay and several other prominent mines, including the North Star near Silverton.

The Crooke brothers owned a smelter in New York City and became the first big time capitalists to invest in the San Juans. They spent a lot of their time and their money refining smelting processes and checking out the local mines for possible investment. They chose Lake City because of its relatively easy access and the (at least initially) rich strikes nearby. The Crooke Smelting Works were on the Lake Fork of the Gunnison River about a mile toward the lake. The Van Geison Works were on the south of Henson Creek. The Van Geison was a lixivation works, which was a leeching method that roasted the ore and used salts to wash out the precious metals. The Van Geison was abandoned by the late 1870s, and it burned in 1900. Mills and smelters were important to any mining town because they concentrated the ores and made low-grade ores worth shipping. Fifty dollar a ton ore might become $250 ore after milling and thereby exceed the cost of shipping ore out of Lake City, which was about sixty dollars to eighty dollars a ton at this time.

Two churches were built in 1876—two of the first on the Western Slope of Colorado. The Presbyterian Church and the Episcopal Church were both started by famous pioneering ministers who spread out over the entire San Juans to preach the gospel. George Darley, minister of the Presbyterian Church, later wrote a wonderful book, *Pioneering in the San Juans*. Darley remembered:

Lake City was a "live mining camp," largely made up of young men of that class who were willing to prospect and take all kinds of chances to make money; but they had no desire to work underneath the ground. Miners were scarce, while prospectors were numerous. No class of men knew better how to treat a minister they liked in a royal manner than the men who went to southwestern Colorado during the great San Juan "excitement" of '75, '76 and '77.

The first public school was started in 1876, paid for mainly by popular subscription. Mr. W. A. McGinnis was the first teacher, and twenty-eight students attended the first session. Colorado became a state in 1876 and some 1,500 people lived in Lake City during the winter of 1876-77, even though thousands more left the area for the winter. The *Silver World* reported, "The amount of freight brought in during 1876 has been carefully estimated and found to be about five million pounds." By 1880 this figure was up to fifteen million pounds-all transported over a hundred miles from Del Norte or Saguache, over a rough, mountain trail.

The flood of new arrivals occurred again in the spring of 1877. Many substantial brick and stone buildings were built that year. In an effort to clean up lawlessness, the town trustees hired four additional marshals and built a new jail. Criminal activity did drop, perhaps because the daytime officers were paid five dollars for each arrest and conviction. Nighttime officers were paid a flat fifty dollars a month for their services.

In April 1877 Lake City had a typical early West gunfight (not like you see on television or in the movies). William Brock and Tom King squared off on Silver Street. They were only eight feet apart when they started shooting. King's gun stuck in his holster, so Brock got off the first shot, which hit King in the stomach. King then shot Brock in the groin and again in the shoulder. King died, and Brock was acquitted. Gunfights like this occurred often, and most times the winner was acquitted, if there was a trial at all. Even if a man was found guilty the sentence was usually mild. Six or eight years seemed to be about average for a killing in a gunfight-usually with parole after just a few years.

A volunteer fire department (the Hough Fire Company) was started, and the first library was opened in 1877. Wells and ditches were dug at strategic locations throughout town to provide a place to hand pump water if needed. The ladies of the community raised the money needed to purchase a large fire bell, and the town later purchased a hook and ladder wagon. A miner's library was also started in 1877. Its stated purpose was "to give unemployed miners something to do." Besides reading material (both books and magazines) it also had stationery, pens, and desks that the miners used to write letters home. It was another part of the ongoing civilizing efforts of the local community. Several large stone buildings and substantial frame residences and commercial buildings were among the 136 new buildings constructed in 1877.

The stage ran daily into Lake City until the heavy winter snows slowed their progress. The stage lines had stations for changing horses every twelve to sixteen miles along the route. Travelers could also wait for the stage and buy tickets at the stations. An agency also existed in Lake City. Large Concord coaches were used, pulled by four horses in the level sections and six horses in the mountains. Stage riders could offer plenty of excitement. Besides the rough roads and steep drop-offs, storms, floods, avalanches, and robbers added interest to trips. Until the railroad came to Silverton, many of the people who were spread out across the San Juans first came into Lake City on the stage, but an equal number walked or rode in on horseback or in wagons. Lake City was described as a "constant stream of humanity pouring into the San Juans through this metropolis." The permanent population of the town and nearby areas was estimated to be between two thousand and two thousand five hundred during 1877.

On September 20, 1877, Susan B. Anthony came to address the town's citizens on women's suffrage at the Hinsdale County courthouse, but so many people turned out for the event that the meeting had to be moved outdoors. Nevertheless, a vote taken after her well-received speech showed that 322 residents thought women should have the right to vote, while 571 were against the idea. In 1877 an official U.S. Land Office was established in Lake City, which allowed homesteaders and unpatented mine owners to get official title (or a patent) to their lands without having to travel long distances. The next closest land office was Durango, so it drew a lot of business to Lake City.

Many new mines were opened and actively operated. Construction of the large Ocean Wave Mill complex was started near the edge of town, and in 1877 the Henson Creek and Uncompahgre Toll Road was opened to Mineral Point. This meant that virtually all the ore from the mines along the route came to Lake City. The Crooke Mill continued to do well and produced $85,498 in silver, $23,698 in copper and lead, and $2,925 in gold during the year. Property values continued to rise, and more than a hundred additional new buildings were built that summer.

In September of 1877 Lake City was the site of a very unusual duel. A man named "Doc" Kaye insulted a man known as "the tailor." They decided to resolve the difference with shotguns. Doc fortified himself with whiskey while the tailor seemed much calmer. They stepped off the distance, fired at each other, and the tailor fell to the ground with a red stain spreading in the snow. Kaye was so horrified that he ran to a nearby pool of water and jumped in. To his amazement the tailor pulled him out. Everyone but Doc knew that the shotguns were loaded with blanks and the red stain was ink. The two men apologized and became good friends.

The year 1878 was a year of a downturn in economic activity for Lake City. It was unfortunately a cycle that would occur many times in the coming years, brought about by rich pockets of ore that stirred excitement, only to find out that they were limited in quantity and contained only lower grade ores. The population of the city dwindled and many of the marginal businesses closed. Helped along by embezzlement of its funds, the First National Bank closed. This bank closure hindered further the investment of capital in the city and its nearby mines. It did, however, help the other two banks in town (the Lake City Bank and the Miners & Merchant's Bank) gain some new customers. Nevertheless, the Lake City Bank was absorbed into the Miners & Merchant's Bank in 1878. The Miners & Merchant's Bank moved into the First National building and continued to operate until 1914.

As bad as things were during 1878, Lake City still held the distinction of having been the top producer in the San Juans up until this point. However, much of the production was a result of having the best smelter in the area (again, credit the Crookes) and the best transportation out of the area (credit Otto Mears). Other areas were either waiting to ship their ores until they had good roads or until an efficient smelter was built that could recover better percentages of their ore with a cheaper process. Although Lake City's future was not totally bleak, it would fall from first to last place in terms of population and ore production among the towns of the San Juans over the next decade.

Although the Utes had ceded the Lake City area in 1873, many of the Utes either didn't approve or didn't understand the treaty. As late as 1879 (shortly after the Meeker Massacre), it was rumored that Ute Chief Colorow and his band had surrounded Lake

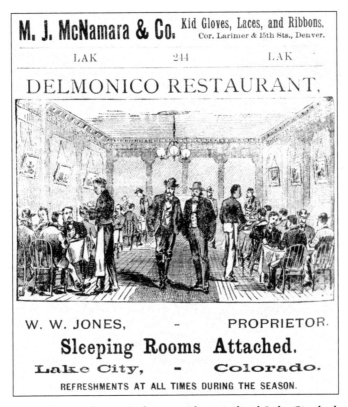

This ad for Delmonico's shows just how civilized Lake City had become by 1884. Courtesy Denver Public Library, Western History Collection. (X77)

City, and the town went on the defense. The citizens formed the thirty-eight member Pitkin Guard and built an armory. Everyone was ready for war, but Chief Ouray calmed the Utes down. The Pitkin Guard never fought the Utes, but they remained organized, bought beautiful uniforms, and formed a drum and bugle corps and a brass band that performed for many years.

Despite the best efforts of the Lake City Fire Department, November 14, 1879, saw a disastrous fire that burned "half of the best block" of commercial buildings to ashes. Almost a million dollars in damage was done to buildings and merchandise. Almost as disastrous as the burned buildings was the loss of a substantial amount of stored supplies that were badly needed for the winter, which led to some pretty tight times before spring thaws allowed more supplies to be brought in.

The Hinsdale County Courthouse was built in the summer of 1879, and the Ocean Wave smelting complex was up and running. The Ocean Wave was pinning its hopes on its mine near Lake City. Unfortunately it didn't pan out. There was good ore (eighty-four dollars per ton) discovered in the Golden Fleece during the winter of 1879-80, but the rich new strike was kept a secret from the general population until the mine had been sold to Louis Weinberg, to allow him more time to file on adjoining claims.

Lake City was a growing and prosperous town in 1882. The smoke in the background is from the Crooke Smelter. Photo Courtesy of Colorado Historical Society. (X7673)

Fossett in 1879 evidently saw the handwriting on the wall. "For some time (Lake City) has been the most populous place in the San Juan country, although Silverton is gaining somewhat at present...the situation is wild and romantic." As if to prove Fossett wrong, mining picked up again in 1880. The Ute-Ulay Mine was at its peak, and the mining camps of Sherman, Burrows Park, and Capitol City had been established. Lake City was now serving a very large area. Eben Olcott, a mining engineer who spent considerable time in both Lake City and Silverton in 1880, wrote that "it is so much more comfortable in Lake City and we have telegraphic lines and much more of a feeling of civilization pervades the place."

In 1881 telephone service was even established between Silverton, Lake City, and Ouray. There was also service to many of the mines and small towns along the way—including Capitol City, Roses's Cabin, Mineral Point, and Animas Forks. Locals celebrated by playing instruments and singing solos, duets, and choruses over the phone lines for hours. Everyone with a telephone was invited to listen in. Several other times during that winter the telephone concerts were held.

Yet with all its "civilization," Lake City still had its felonious side. The big trouble wasn't so much in town as on the way to town. The stage from Del Norte to Lake City was robbed five times during late 1880 and early 1881. The robbers were probably after money that might be found in the mail, but they were also robbing the passengers. Several of the stage passengers were wounded, and many Lake City residents lost money coming by mail that they badly needed. Finally, in May 1881, Sheriff L. M. Armstrong and James P. Galloway captured Billy Le Roy and his brother, who used the alias of Frank Clark. They were taken to the Del Norte jail, but while the sheriff slept, a mob overcame him, took the LeRoys, and hung them from a cottonwood along the Rio Grande that same evening. The lifeless bodies were then returned to the jail in the morning. The stage robberies stopped!

Ever optimistic Frank Fossett, in 1880, reported on Lake City:

Hinsdale County is the most easterly of the important silver districts in the San Juans. Its metropolis is Lake City dating from 1874-75 located at the junction of Henson Creek and Lake Fork of the Gunnison....There are numerous silver lodes in the lofty mountains that rise almost perpendicularly for a half mile on every side-many of them worked extensively...The site is decidedly romantic, surrounded as it is by stependous (sic) mountains....From a mere cluster of cabins in 1875 it has grown into a thriving busy center of from 1,500 to 2,000 population, its mills and reduction works comprising the most extensive system of mining machinery in all the San Juan country. It has churches of almost every denomination, three or four hotels, good schools, several banks, five saw mills, free reading room and library, two excellent and energetic newspapers and other evidence too numerous to mention of substantial and lasting prosperity.

However the 1880 census reported only 1,487 persons living in the entire county, and the fall of 1880 was full of rumors of Ute uprisings. But typical of the cyclical nature of Lake City mining there were new strikes of rich minerals at Palmetto Gulch near the top of Engineer Pass that same year, as well as in the vicinity of Capitol City and at the Golden Wonder near the Packer Massacre site in Deadman's Gulch. The second major rush to Lake City occurred between the fall of 1880 and 1882. The Crooke Smelter (processing Ute-Ulay and Polar Star ore) and the Palmetto Mill up above Rose's Cabin were full to capacity. Ingersoll reported "over five million pounds of mining machinery and supplies were taken in by wagons during 1880, at a cost of over a million dollars for transportation alone.... At the beginning of 1881 about 2,000 people lived in the town itself, not counting the great number of men in the mountains round about." Lake City was large and prosperous but there was one striking difference in the way the town looked then compared with today-there were no trees! They had all been cut down for construction.

Although not a good reproduction, this photo shows the old Ocean Wave bridge and shows why it would be a good place for a hanging. Courtesy Denver Public Library. (X72)

This view of Silver Street is looking south in 1894. The two story brick building near the center is the bank. The veranda on the left belongs to the Pueblo House. Courtesy Denver Public Library. (X73)

Lake City also had the usual red-light district—this one was located on the upper or south end of Bluff Street (the westernmost street in town) close to Henson Creek. It was usually referred to as "Hell's Acres." One of the most famous madams on Bluff Street was Clara Ogden, who arrived in Lake City in the late 1880s. She built the Crystal Palace dance hall, which was by far the largest and grandest dance hall and bordello in Lake City. On the ground floor were parlors and a ballroom furnished with crystal, mahogany, mirrors, and thick carpeting. It included a second story that contained scores of bedrooms. Clara also believed strongly in advertising. She was known to fill her elegant carriage with girls and drive out to the nearby mining camps to show off their wares. It was said that she hoped to open a chain of bordellos in the outlying areas, but she never did. Clara was forced to close down near the turn of the century after there was a shooting in her establishment.

Although generally a law-abiding town, in 1882 Sheriff Edward N. Campbell was killed while trying to stop a burglary. Deputy Claire Smith, who was also on the scene, recognized the killers as George Betts and James Browning, proprietors of the notorious San Juan Central Dance Hall. The men were captured and jailed, but irate citizens immediately began talking about a lynching. Shortly after midnight a large number of armed and masked men moved on the jail. Although the law officers tried to reason with the mob, the men broke off the jail's lock with a sledgehammer, dragged the men from their cells, and hanged them from the Ocean Wave Bridge. School was even let out later that day so the children could come view the bodies. The two were later buried in the Lake City cemetery. George Betts, the more notorious of the killers, didn't give up his soul easily-many people near the bridge have reported seeing his ghastly looking ghost wandering near where his corpse had been left hanging. James Browning's ghost has also been reported to have appeared at the courthouse as if awaiting a trial. This

This disastrous June 15, 1901, fire leveled a good bit of Lake City's business district. Courtesy Colorado Historical Society. (X7666)

and many other ghost stories from the San Juans are well told by Mary Joy Martin in *Something in the Wind* and *Twilight Dwellers.*

By 1884 the cycle had reversed again. Ingersoll reported, "Lake City is not now so active as formerly." The problem was that although there were large quantities of ore, it was generally too low-grade and not worth shipping. Lake City was also too far from a railroad. Sapinero at present-day Blue Mesa Lake was the closest railhead. When the Crooke Mining and Smelter Company closed its mines (including the Ute-Ulay) and smelters in 1883, the value of silver and lead actually processed and purchased at Lake City fell from $208,703 to $11,362 in a year. When the Ute-Ulay reopened in 1887 silver and lead production went back up to $113,000, but Lake City would never be the same.

Croffut reported in 1885 that:

The stranger visiting here will be surprised to see the great number of stores, hotels, livery stables, saloons and shops of all kinds, all of which seem to be doing an unusual amount of business for the size of the place. The explanation can be found in the fact that the city is located in the center of a score or more of small mining camps, numbering all along up to 300 population each. The people from the geographical location of the city, find it the best and most convenient place to purchase their supplies, spend their money, and sojourn for a season of recreation.

Although Lake City mining had been in a severe slump for some while, in 1889 the Denver & Rio Grande Railroad arrived, and the town was for the first time connected to the outside world by cheap transportation. Another mining boom occurred since many of the area mines were able to reopen and ship their low-grade ore out at a profit. At the very same time large amounts of high-grade silver were found at the Ute-Ulay Mine and large and extremely rich deposits of gold were found at the Golden Fleece. With this new-found prosperity the first city water works was completed in 1890 and an electric power plant was built in 1891. That same year a city-wide electrical system was installed. By 1891 the city's population had rebounded from a low below one thousand to about two thousand. It

The Hinsdale County Courthouse, built in 1877, is most famous for being the site of the Alfred Packer murder trial. Courtesy Denver Public Library. (X71)

was a time of general prosperity-a time when some of the run-down properties around the town were fixed up, the Lake City Brass Band was established, and many new houses were built. In 1890 there were twenty mines shipping ore. In 1891 the Ute-Ulay alone shipped four hundred thousand dollars in ore. From 1892 to 1902 more than a half million dollars in ore was shipped yearly—not a great amount compared with the other San Juan cities but substantial.

But all was not rosy. In the fall of 1891 the influenza epidemic hit Lake City. More than a hundred persons died that winter. The city fathers even resorted to turning on all the fire hydrants in town and leaving them open in an apparent attempt to wash away the germs of the epidemic. As bad as the epidemic was, it wasn't as bad as in Silverton.

Lake City was also hard hit by the Silver Panic of 1893, as were all the mines and towns of the San Juans. Silver production was at an all-time high in the Lake City area when the bottom fell out of the silver market; but just as the economy looked like it would be the worst in Hinsdale County history, gold production tripled during 1895. The gold boom was led by even richer strikes at the Golden Fleece, and in 1897 the Hidden Treasure shipped its first rich ore. Lead and silver were also being found in much larger quantities. However, the boom and bust cycle continued, and, by 1903, produc-tion was slowing down again — even the gold strikes were fading.

With the beautiful mountains and pristine Lake San Cristobal nearby, tourism started establishing itself early at Lake City. Hall wrote in his 1895 History of Colorado "the visitor is lost in wonder at the variety and general magnificence of the scenery, the fantastic rock formations, the marvelously picturesque contour of the ranges on either side, and the loveliness of the entire valley."

However even in the twentieth century Lake City still had its Wild West flavor. On April 1, 1901, John Addington and Alexander Surtees had a gunfight over a girl. They met on Silver Street between the bank and the Hough Building and they both fired and struck each other. Surtees died immediately, and Addington passed away two days later. Dr. Benjamin Cummings, who ran into the street when he heard shots, was hit in the

A decorative iron fence surrounds a family plot in the old Lake City Cemetery. Courtesy Denver Public Library. (X7670)

buttocks and had to find another doctor to operate on him. The bullet wasn't found in his flesh, but later when the doctor pulled his silk handkerchief from his rear pocket, the bullet fell out.

The local mines were producing the least amount of ore of any county in the San Juans. As the mines closed; people left town. Lake City's population dropped drastically over the future decades. The census gave Lake City a year round population of 1,000 in 1900; but by 1910 it was 405, in 1940 down to 185, and in 1970 only 91. However, the summer population was always much higher.

In 1933 the D&RG Railroad abandoned its branch line into Lake City. Between 1938 and 1946 there was no newspaper operating in Lake City. It was a dark time- many felt the town would never survive but it did, and today it is a popular summer tourist attraction. And although just a fraction of what it was in the 1870s and 1880s, there has always been some mineral production around Lake City, right up to the present.

Many wonderful old buildings in Lake City are well worth seeing. The wooden Hinsdale County Courthouse was built at the corner of Gunnison and Third in 1877. The second floor courtroom was the site of the infamous Alferd Packer trial. Since the courthouse held the largest room in town (thirty five by forty five feet), it also became a meeting place for socials, church gatherings, and meetings. The courthouse is the largest wooden building left in Lake City. The original 1893 Lake City jail burned down but the steel cages that were installed inside it are now on display at the city park, along with an original D&RG narrow gauge caboose.

The Hinsdale County Museum at 130 Silver Street is well worth a visit. The building was built in 1877 as the Henry Finley Block at a cost of $8,225. It was originally a general merchandise store, then a hardware store, grocery store, and, after 1900, a saloon and the lodge of the Independent Order of Odd Fellows (IOOF). The Hinsdale County Historical Society bought the building for a museum in 1989, and the well-restored Smith-Grantham house was moved next door by the historical society in 1993.

Lake City is famous for its old log cabins and beautiful Victorian homes that can be seen throughout the town. There are too many residences to describe here, but Grant Houston has written *Historic Homes of Lake City*, which does a great job of detailing their history. Many of these homes were built in the late 1870s or the late 1890s—both periods of boom. A large part of the original business district burned, but many extremely well-preserved old buildings still remain.

The Lake City cemeteries are well worth a visit. There is the City Cemetery and the IOOF cemetery. The IOOF cemetery is off Balsam Drive just outside the north end of

The Lake City Sunday School, parents and teachers, pose outside the Baptist Church in November 1898. Courtesy Colorado Historical Society. (X7670)

town. It is often called "the new cemetery" to differentiate it from the City Cemetery, which was established in 1876. The IOOF was established only one year later in 1877, but it was not opened to the public until a number of years later. As with all the San Juan graveyards, the gravestones tell the story of the town's prostitutes, the all too common deaths of young infants or women in childbirth, and deaths caused by gunfights and epidemics.

The Armory Building, which is now one of the premier spots in town, is located at Bluff and Third Streets. It was built in response to rumors of the Ute uprising in 1879, which led to the formation of the Pitkin Guards for whom the building was built. The building became a focal point for community affairs and was later used as an opera house. Now it is used as a gym and a meeting place for the larger gatherings that are held in the town.

The Presbyterian Church (located at the southwest corner of Fifth Street and Gunnison) held its first services on June 17, 1876. One hundred fifteen people attended and agreed to immediately start construction of a building. They dedicated their church building on November 19, 1876, just one year after the founding of the town. The bell for the church was made in New Jersey and brought in from Pueblo in 1877. The church's reed organ was bought in 1882 and is still used in the church. The Presbyterian

Church was founded by Alexander and George Darley in 1876 and is the oldest church on the Western Slope. Darley was the Presbyterian minister in Del Norte and the missionary minister to the San Juan area. Before the church building was built, Rev. Darley would preach in the local saloons, the owners stopping their gambling games for the services. The church immediately became not only a religious center but also a social center for the people of Lake City. Darley remained as minister until 1880. The Presbyterian manse is located next to the church.

St. James Episcopal Church (located on the northwest corner of Fifth Street and Gunnison) was organized in January 1877, and its little building erected in March, 1877. Parson Hogue fought his way over the mountains to preach in Lake City, like Rev. Darley and originally held his services in the saloons. He was known to sometimes take money from the offering plate to the local gambling halls in an attempt to increase "the Lord's share." St. James was the first Episcopal Church on the Western Slope but had no regular priest until May of 1880. The first baptism at the church was a child born in Silverton and brought sixty-five miles on horseback for the ceremony in Lake City.

Father Hayes from Del Norte held the first Catholic mass in Lake City in September 1877. The Roman Catholic Church of St. Rose of Lima was dedicated (although only partially finished) on January 6, 1878. By March the building was complete. It is located on the bench where Henson Creek flows into town. The Catholics, along with the other churches in town, frequently held social events—parties, plays, benefits, literary programs, and much more.

The first Baptist service was conducted in a house in Lake City on July 9, 1876. The first regular services were held in September 1883. The magnificent Victorian-style Baptist church was built in 1891 on Bluff Street, extremely close to the activities going on at Hell's Acre. The first service was September 20, 1891, and the church was dedicated on January 17, 1892. Like today, there were always a large number of children involved in the church's activities.

The two-story Lake City School was built in 1880, but only the first floor was occupied until 1893. It was remodeled in 1949 and remodeled very extensively again in 2002 to its present appearance. A Mrs. Gage was the first teacher,

Today, Lake City is a charming town with large cottonwoods, old Victorian homes and log cabins lining its wide streets. For a time Hinsdale County was reported to have the smallest population of any county in the United States, but it has grown again in recent years; and percentage-wise is one of the fastest growing counties in the United States. (Most of the counties in the San Juans also hold the same distinction as people discover our beautiful mountains.) Let us hope the days of boom and bust cycles are a thing of the past for Lake City and that the future continues to look continuously rosy.

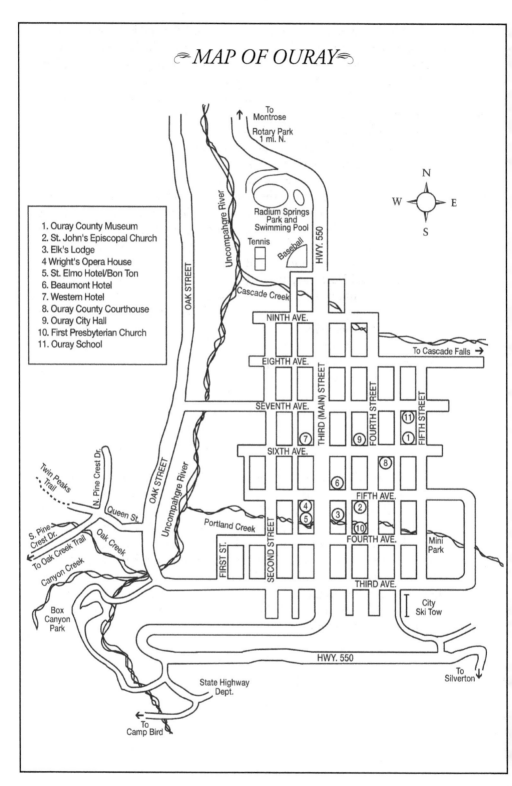

MAP OF OURAY

1. Ouray County Museum
2. St. John's Episcopal Church
3. Elk's Lodge
4 Wright's Opera House
5. St. Elmo Hotel/Bon Ton
6. Beaumont Hotel
7. Western Hotel
8. Ouray County Courthouse
9. Ouray City Hall
10. First Presbyterian Church
11. Ouray School

⪻ Chapter Four ⪼
THE CITY OF OURAY

The Ute Indians arrived in the San Juans about 1300 A.D., and it is quite likely that they soon discovered the hot springs that surface in the beautiful valley below the amphitheater that now holds the City of Ouray (elevation 7,705 feet). Trappers certainly made their way into the bowl in search of beaver as early as 1820, and there is evidence that prospectors may have visited the site during 1849 and 1850 on their way to search for gold in California.

When the first gold strikes were made in Colorado in 1859, the new arrivals soon spread out over the nearby mountains, and some of them must have found their way into the little bowl that now holds the City of Ouray. A diary exists of a small group of men who wintered just a few miles to the north of Ouray at the mouth of Coal Creek in 1860. George Howard, a member of the Baker party, many years later said that a portion of the Baker party came into the San Juans from near present-day Ridgway where they found placer gold. The discovery in 1875 of an old weathered wagon with mining tools hidden underneath seems to verify this story. Howard said this group went up the valley through Ouray and then over to Baker's Park. It is likely that this happened in 1861.

There is an account of Charles Brugh, who prospected the Uncompahgre River in 1861 and then attempted to go to Baker's Park. His group ran out of provisions in heavy snowfields, and Brugh later claimed that he was sure his companions were going to eat him, so he took off alone. His mind may have played tricks on him due to his condition, for when members of Baker's group found him wandering deliriously near Engineer Mountain, he weighed only forty-eight pounds!

There is also a lost gold mine legend involving the Ouray area. Two prospectors supposedly made a rich gold strike in 1863 along Oak Creek just above present-day Ouray. They stayed there for several months, breaking the ore out by building fires and then throwing cold water on the hot rock. Eventually they fled in fear of the nearby Utes, leaving much of their gold behind. The Civil War was raging and the area around present-day Ouray was deserted for almost a decade.

Finally in 1875, almost two years after the San Juans were ceded by the Utes, large groups of prospectors began to make their way down both the Uncompahgre River and Bear Creek from Mineral Point. In July of 1875 A. W. "Gus" Begole and John Eckles entered the heavily wooded bowl and made promising discoveries of rich minerals. They went back to Silverton to get more supplies and returned with other men. On August 11, 1875, Jacob Ohlwiler, Begole, and John Morrow located the Cedar and Clipper lodes within the present-day Ouray town site. On August 23 A. J. Staley and Logan Whitlock discovered the Trout and Fisherman lodes, which were also within the present-day city limits. The mining claims were so named because the two men were fishing for trout at the time of their discoveries.

It was not rich minerals, however, that made the site of Ouray so valuable. It was perfectly suited for a town site, as it was located at a relatively low elevation at the

Chief Ouray and his wife, Chipeta, were known for their peacekeeping efforts. Through the use of treaties, Ouray was able to hang onto large parts of Colorado for over twenty years. Courtesy Denver Public Library, Western History Dept. (F1685)

end of a broad valley and yet was near many of the newly discovered mines. Eight prospectors, therefore, filed a declaration notice for the town of "Uncompahgre" on August 28, 1875. Then in October 1875 Begole and Eckles discovered the rich Mineral Farm Mine just a mile southwest of the new town. There were enough mining prospects in the area that several dozen people decided to spend the winter of 1875-76 in Uncompahgre City.

Log cabins were built, wagons were sent for supplies, and several men were able to work their claims all winter. One woman spent the winter at the new camp, during which time she gave birth to a child. Ouray's first post office was established on October 28, 1875, but it was little more than a promise to have the mail brought up from the Los Pinos II Ute Indian Agency. Otto Mears had received the contract to bring in the mail to the agency. When the snow got extremely deep that winter he resorted to dogsleds and then eventually carried the mail on his back while using snowshoes. Although spring came early in 1876, the Uncompahgre residents almost died of starvation that winter.

Unknown to those in Uncompahgre City, there were many more men in Imogene and Yankee Boy Basins just a few miles to the southwest of the town, but steep canyons and sheer cliffs blocked communications between the two areas. Andy Richardson, who named Imogene Basin after his wife, and his partner, David P. Quinn, prospected most of the basins in and around Imogene and Yankee Boy in 1875 and 1876. Richardson spent the winter near Yankee Boy Basin, and his cabin became the meeting place for the prospectors in the area. Yet the men at Uncompahgre were by no means isolated. Just twenty easily traveled miles to the north was the second Los Pinos Ute Agency. There white men, supplies, and mail could be found. Certainly some of the residents of 1875 visited the agency often, as Chief Ouray was friendly toward most white people.

Ouray had acted as an interpreter at the Conejos Peace Treaty conference in 1863, at which time the Utes entered into a treaty gave up most of that part of present-day Colorado located east of the San Juans. Ouray translated from Ute into Spanish, a language in which he was fluent since he had grown up near the Taos, New Mexico area. Ouray's Spanish was in turn translated into English. After the Conejos Treaty it was decided by the whites that Ouray should become the first overall Ute Chief since the U.S. government found it too cumbersome to try to deal with the dozens of Ute chiefs that actually existed. Somehow through sheer force of personality, Ouray actually gained general control over the many Ute tribes.

A 1905 pamphlet summed up the contemporary white man's feelings toward Ouray:

Prior to 1881 the Ute Indians held undisputed possession of the Uncompahgre Valley. The head of all their tribes was Chief Ouray, who for many years swayed a scepter of such equity that he was known as the "white man's friend." Never in all his dealings with the whites did he show himself other than their friend, tried and true. Once, while in council with the whites upon some important matter, one of the under chiefs arose and began a tirade against the white man; old Chief Ouray listened for a moment, and with the fire flashing from his eyes he arose leveling a six shooter at the turbulent speaker, uttered one word "hikee," which is Ute for "get out." The under chief well understood the meaning of the command and scrambled for cover, when the meeting was continued to its conclusion without further interruption.

It was natural that the residents of Uncompahgre City sought the protection of the great chief. In fact, when the town was surveyed, laid out, and incorporated October 2, 1876, it was named "Ouray" in his honor; but the original Uncompahgre plat was used for the new town. The details of Chief Ouray's life can be found in *Ouray: Chief of the Utes* by P. David Smith.

Today many visitors want to know the meaning and correct pronunciation of the town's name. The chief himself said that his name had no Ute meaning and was merely a sound (oo-ay) that he made constantly as a child. However, some historians report that the name means "the arrow." The Utes have trouble pronouncing "r" so it is probable that the name was pronounced "Ooay" or "Olay." Today the name is pronounced "Ouray" as rhymes with "Hooray."

Those hardy individuals who spent the winter of 1875-76 in the small settlement of Ouray suffered many hardships, but there was one goal that kept them going—to be the first on the scene in the spring of 1876. They knew a rush would come, and it did. In the early summer of 1876 Mr. and Mrs. Dixon started the first hotel in their

This extremely early view of Ouray was taken from a stereo card made about 1878. Most of the trees have not yet been cut down and even Main Street is just a rough path. Author's Collection.

log cabin, and Mrs. Dixon also furnished meals to the locals. Guests had to sleep on the floor and bring their own bedrolls, but the Dixons did well enough that they built a two-story log cabin the next year.

So many people came to the new town of Ouray that by the winter of 1876 there were over 400 inhabitants, 214 cabins and tents, a school, four stores, two blacksmiths, two hotels, a sawmill, an ore sampling works, a post office, and the requisite number of saloons, gambling houses, and prostitutes to support such a population. In 1877 the Barlow and Sanderson stage started regular operations into Ouray from Saguache. The trip went through Lake City and the Los Pinos II Agency and took thirty-four hours. The toll for travel over the road was extremely high—twenty dollars for a six-mule team wagon.

Three different preachers arrived in Ouray in 1876. Originally they preached at some of the local saloons. The Rev. George Darley was the first minister to reach Ouray, and he also built the first church. The Baptist and Methodist churches also had a presence. Parson Hogue established the Episcopal Church in 1876. The Episcopal Church building is the oldest church building still standing in Ouray.

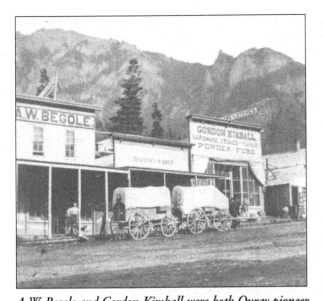

A. W. Begole and Gordon Kimball were both Ouray pioneer prospectors who took their money from selling mining claims and put it into stores in Ouray. Courtesy Denver Public Library, Western History Dept. (F28385)

Although the original founding of Ouray was based on the mining industry, it wasn't long before it also became a tourist attraction. The nicknames "Opal of the Mountains" and "Gem of the Rockies" were applied to Ouray early in its existence; and if Ouray was a gem, it had a setting to match. Three major waterfalls are within the city limits or just a short distance away. Five creeks flow into the Uncompahgre River within the city limits. The amphitheater encircles the city to the east, and high mountains (the east end of the majestic Sneffles Range) rise directly to the west of the city. So fully do the mountains as well as the maroon and yellow cliffs surround the town that the visitor who arrives at night often wonders how to get out the next day.

Within just a few years of the arrival of its first permanent residents, Ouray enjoyed a fair amount of tourists. Fossett in 1879 wrote, "few towns in the world are so beautifully located as Ouray.... Grand and majestic scenery, health-giving mineral waters, and some of the best silver mines in the state are some of the attractions... Population 700."

The arrival of the first sawmill in 1877 meant that frame buildings could be constructed. It also meant that most of the trees in the beautiful little bowl were soon cut down. When Ouray County was established from the northern part of San Juan County on January 18, 1877, the City of Ouray was made the county seat. After elections, the first county commissioners meeting occurred March 7, 1877. Jess Benton (later to become sheriff) built a two-story frame building called Benton Hall that eventually became the town's combination courthouse, city hall, meeting hall, and church.

Ouray's first school was started on June 18, 1877. Miss Libbie King was the first teacher. The town was already crowded with children—forty-three attended the first session. Ouray built its first water works in 1877. In 1877 Ouray's first bank (The Ouray Bank) was established. The Thatcher brothers of Pueblo formed the Miners and Merchants Bank in 1878. After the Beaumont Hotel was completed, the bank was moved to the southwest corner of that structure. The Thatcher brothers put up much of the needed capital for the mines around Ouray including the $600,000 needed to blast out the Revenue Tunnel at Sneffles. Ouray's first newspaper, *The Ouray Times*, was first published in June1877. In 1879 the *Solid Muldoon* arrived and offered major competition. Publisher David Day's wit was known as far away as Europe.

The town made a huge (thirteen by twenty foot) flag to celebrate the Fourth of July in 1877, and it was placed on a fifty-foot flagpole in the middle of Main Street where it flew for many years. By 1878 Ouray's population had risen to almost 800, and by 1880 Ouray had 864 citizens, of whom 84 were of school age. But the town still had a precarious existence. It could easily have disappeared overnight, if the mines around Ouray hadn't been doing well. The Mineral Farm was sold in 1878 for seventy-five thousand dollars. The Virginius was sold for one hundred thousand dollars in 1880. The Wheel of Fortune had sold for one hundred sixty thousand dollars in 1877, and by 1882 hand-sorted Yankee Boy ore was averaging 1,231 ounces of silver per ton.

During the last half of the 1870s most mining activity in the vicinity of Ouray was centered along Canyon Creek up to Yankee Boy and Imogene Basins. By the late 1870s there were three small smelters within a few miles of town. There were no good roads to the mining prospects and eventually Otto Mears was called in to connect the mines with the town. At the beginning of the 1880s Ouray was still in a precarious position, even though it had grown to one of the top ten largest towns in Colorado.

As in the other San Juan towns, there were killings in Ouray, but a number of them involved the local sheriff. Sheriff Jesse Benton killed a man named Lucas in front of the Dixon House Hotel. Benton also shot and killed a Chinaman who was accused of an attack on a white girl. Ed Leggett was killed in a dance hall fight. Benton killed enough men that there were serious questions as to whether his motto was "shoot first and ask questions later."

In September 1881 the Utes were forced out of Colorado to northeast Utah. They obviously didn't even understand they had agreed to leave and many only left at gunpoint. Yet for the first time the San Juans could be approached through non-Indian land from the north. It had been legal for whites to travel through the Ute Reservation,

Ouray existed as a supply town for the many rich nearby mines, most of which were only accessible by pack mules or burro trains. W.H. Jackson Photo. Author's Collection.

but most had been too afraid to do so. The Denver & Rio Grande Railroad was immediately built toward present-day Montrose, arriving there in 1882. Although the main line went on into Utah, a branch headed toward Ouray; however, due to various circumstances, it took five years to finish the job.

In 1882 the rich ores of Red Mountain were discovered half way between Silverton and Ouray—almost straddling the county line. The first large discoveries were in San Juan County, but eventually it was realized that most of the rich ore was in Ouray County. Otto Mears, together with the County of Ouray, built the Million Dollar Highway in 1883 to connect with new mines.

Many of the people who rushed into the Red Mountain District came down to Ouray for their trips to the "big city," and many of the Ouray merchants sent supplies to Red Mountain. Ouray progressed from a mining camp to an actual town during this period. It's future seemed assured.

One of the few hangings of a woman occurred in Ouray on January 24, 1884. Michael and Margaret Cuddigan had mistreated their ten year-old adopted daughter so badly that she died. The Cuddigans were investigated after the death and arrested. The child's body was examined and it was found that when she died her fingers, feet, and legs had been badly frozen. She also had many bruises, was suffering from malnutrition, and may have been raped. The couple was taken from the jail and hung from a tree that used to stand at the ballpark near the pool. When the people of Denver criticized the citizens of Ouray for their actions, they sent the little girl's body to the Colorado capitol city where it was looked at by thousands of people. Everyone agreed that the citizens of Ouray had done the right thing.

In September 1887, Joseph Dixon, who was employed by the Delmonico Restaurant in Silverton, was in Ouray. Dixon had worked with twenty-nine year old Nellie Day

The Dunbarton Plunge (left foreground) was located at the same spot as today's Weisbaden Lodge and Hotel. The city's new two-story brick school can be seen in the background. Author's Collection.

at the Beaumont Hotel. They had a fight that ended with Dixon shooting Day at close range, killing her instantly. He was immediately arrested and taken to jail. That night a mob went to the jail and tried to break in. When they were unsuccessful they wrapped the wooden jail in blankets, soaked them with kerosene, and set fire to the building. After the fire was put out Dixon was found burned to death in his cell.

The Denver & Rio Grande Railroad was finally built up the west side of the Uncompahgre River to Ouray in 1887. Crofutt had written in 1885 " if the miners of Ouray pray at all, it is for the coming of the 'Iron Horse.'" The railroad chose this unusual route on the cliffs because David Day, editor of the *Solid Muldoon*, tried to make a financial killing by buying the land logically needed by the railroad for the route on the east side of the river. The City of Ouray responded by helping pay for the cost of extending the line on the west side of the river, leaving Day high and dry and hated by most of the Ouray merchants.

In 1888 the Silverton Railroad was built from Silverton to the Red Mountains and after that most of the ore went out through Silverton. The Guston, Yankee Girl, and National Belle were some of the big mines. Several thousand people lived in five major communities on Red Mountain with Ouray and Silverton competing for the district's business.

Ouray continued to thrive as a tourist attraction. Ouray's first hot springs swimming pool was built at the Radium Springs Court on Fifth Street in the 1880s. It is the present site of the Wiesbaden Motel. The late 1880s and early 1890s were a time when tourists began to really discover the San Juan's beauty. Early travel writers continued to extol Ouray's charms and came to soak in its hot springs. Box Canyon, the Fern Grottos of the Uncompahgre Gorge, and the Crystal Caves were all popular attractions. From the first day it was opened the Million Dollar Highway from Silverton to Ouray was touted as a chance to visit "God's Country." The D&RG promoted travel across the road by stagecoach as late as the 1920s when it was promoted as "a ride through the past." Travel writer Ernest Ingersoll wrote these words in 1885—less than ten years after the town of Ouray had been founded in the heart of the rugged San Juan Mountains:

> Ouray is—what shall I say? The prettiest mountain town in Colorado? That wouldn't do. A dozen other places would deny it.... Yet that it is among the most attractive in situation, in climate, in appearance, and in the society it affords, there can be no doubt. There are few western villages that can boast so much civilization.

That same year Ingersoll wrote "Ouray's principal claim to our notice as sightseers lay in its beautiful situation—a collection of pictures which it would be hard to duplicate in an equally limited space any where in the whole Rocky Mountains."

The effect on mining near Ouray by the arrival of the Denver & Rio Grande Railroad was extraordinary. Imogene, Yankee Boy, Red Mountain, and even Telluride ores, could be shipped much more economically. Ouray truly boomed and many of Ouray's finest buildings were built during this time. The population swelled from 864 in 1880 to 2,534 in 1890. Wages were good, but Dave Day reported that there was "no demand for preachers, lawyers, book agents, tramps, or ornamental nuisances."

By 1890 Ouray was taking on an air of real prosperity. Since the census of that year showed the town to have over the necessary two thousand persons, Ouray officially

The Ouray Hose Cart Team had their annual fund raising banquet at the Dixon House Hotel in 1885. It took too much time to hook animals to the hose cart so men pulled it. Author's Collection.

became a city instead of a town. The Beaumont Hotel had been completed in 1887. Ouray's school system took a major leap forward in 1888 when a new two-story, four-room brick schoolhouse was built. It was replaced by a new building in 1938, which is still used today, although remodeled extensively several times. The streets had been graded and wooden sidewalks built. Ouray's electric plant had been installed in 1885.

The city officials wrote their own ordinances in 1891. Most provided for practical health and safety issues of the times. Small pox and typhoid laws indicated those were dreaded problems. But there were also laws that pistols, daggers, and other deadly weapons could only be carried for "legitimate" purposes, and that children under sixteen could not be in houses of prostitution or where obscene plays were performed "unless able to give a lawful excuse therefore." Women could not be bartenders, bikes could not be ridden over eight miles an hour, and children under eighteen had an eight p.m. curfew, although the legal drinking age was sixteen.

Part of Ouray's new air of prominence was due to actions that had been taken to fight fires. The volunteer fire department was established in the early 1880s and by 1889 Ouray had both a hook and ladder and a hose company—each with its own uniform and special benefit dances to help them raise money. Another fire-fighting practice was the use of brick and stone construction. Bricks were turned out by the thousands from a spot near the present-day swimming pool. The hospital, school, courthouse, Beaumont Hotel, City Hall, and many large commercial buildings (like the Wright Opera House) were the result.

Along with culture and prosperity came the seedier side of life. There were as many as thirty saloons, gambling halls were everywhere, dance halls were popular, and houses of prostitution lined Second Street between Seventh and Eighth Avenues. The Temple of Music, Bon Ton, Bird Cage, Clipper, and Monte Carlo were just a few of the bigger operations. Over a hundred girls worked at the establishments. Probably the biggest was John Vanoli's Roma Saloon, 220 Bar and Gold Belt Theatre complex on Main Street between Eighth and Ninth Avenues. The Roma was rowdy; Vanoli himself committed one of the murders at the establishment. He went to prison for only eight months before being pardoned. What was the reason for the short sentence? Most felt that his victim was a bad man and deserved to be killed! It didn't seem to matter that Vanoli shot first, hit his victim, and then shot him again while he was laying on the floor! In 1895 Vanoli got into another mess at the Gold Belt. Even though he shot his victim three times (one of which was in the back) no charges were brought. However, the event sparked a movement to close down dance halls, gambling establishments, and bawdy houses in Ouray. Saloons were even required to close at midnight and on Sundays.

Ouray has always had a large 4th of July celebration. This parade includes the militia formed for protection from the Utes and the hose cart company above them. The large flag is on the flagpole and the Beaumont has been built, so it was about 1890. Courtesy Colorado Historical Society.

Ouray was reported to have as many as thirty-five saloons during the late 1890s. Besides the Roma there was the Free Coinage, Corner Saloon, Bucket of Blood, and the Bank (so named because a patron could tell his wife he was going to the bank and not be lying). During this time Dave Day noted, "The ladies of the Episcopal Church have organized for their summer campaign against sinners. They will devote their energies to rescuing miners and editors (Day did spend a lot of time in the bars) as experience has taught them that lawyers and bank cashiers are not worth saving." During a period of just twenty days the flu epidemic killed sixty-seven persons in Ouray in the fall of 1891. Smallpox and typhoid fever were also major problems during the 1890s.

The 1893 Silver Panic hurt Ouray as well as all the San Juan towns. Times got so tough that the city council turned off the electric street lights, stopped working on the streets, and cut the city employees' salaries. However, gold had been found in 1889 in the nearby mines at Gold Hill. The American Nettie was perhaps the most famous of these mines.   It was connected to its mill on the valley floor by a 6,300-foot tram. The weighted buckets going down pulled the empty buckets back up. The American Nettie gold ore was found in rich pockets of iron and copper pyrites, galena, and grey copper, and the mine eventually produced over two million dollars in ore.

On the backside of the mountain (called Gold Hill) was the Bachelor Mine, which was also tied in with the Wedge and Khedive. Its ore was a very rich zinc-lead-silver, which was worth mining even with depressed silver prices. Eventually it produced over three million dollars in ore. The small town of Ash was built at the site of the Bachelor Mine on Canyon Creek (originally called "Red Canyon"). Ash had a peak population of about one hundred in 1900, but nothing remains of it today.

The 1893 Silver Crash not only closed many of the San Juan silver mines, but Ouray's first bank (The Bank of Ouray) was also forced to close. However, some of the Red Mountain mines were so rich that they continued to produce silver, even though they received only half the previous price. It also helped that many of the local ores had always carried small amounts of gold; and base metals like copper, zinc, and lead were becoming more valuable. By 1895 the Ouray economy was again stabilizing; and when large amounts of gold were discovered at the Camp Bird Mine in 1896, Ouray's continued existence was assured.

Because of the Silver Crash of 1893, the population of Ouray had fallen to 2,196 in 1900, but that drop in population was minor compared to many Colorado mining towns. However, over the long haul, metal prices continued to slide. By 1918 even Camp Bird production was falling, although the mine would net an additional twenty million dollars before its final closure. Tourism continued to become more important. Box Canyon was developed. The Ouray Hot Springs Pool was built in 1927. The Million Dollar Highway not only brought people to Ouray, but also became an attraction in itself. In the 1960s Ouray came to be known as "The Jeep Capital of the World." By 1980 mining was basically dead, but Ouray continued to thrive as a tourist attraction.

Today the amphitheater above Ouray contains a Forest Service campground (open only in the summer) as well as a large network of hiking trails collectively referred to as the Portland Trail. It can be reached from U. S. 550 just south of the city. The terrain in the amphitheater is some of the easiest around but is still challenging and beautiful. The two to three hour hike to the Chief Ouray Mine is popular, but the trail is steep and contains about a dozen grueling switchbacks before sweeping to the north. The mine is located at 10,000 feet and can be seen above Cascade Falls from many parts of town. The amphitheater also contains the Grizzly Bear Mine—a 6,000-foot tunnel that accesses rich silver ore underneath Bear Creek. It is one of the few mines being worked in the San Juans at this writing, although on a very limited scale. The amphitheater also provides excellent cross-country skiing in the winter. Deer, elk, mountain sheep, and an occasional bear can be seen in the area.

On the southwest side of Ouray is an unusual geologic feature—Box Canyon Falls. The waters of Canyon Creek have worn a narrow gorge through the ages, where the waters descend in an unending roar that shakes the earth. In Ouray's early days tourists would walk up the bank of the creek and look at the falls, but they could not enter the canyon itself. Now the City of Ouray permits visitors (for a small charge) to walk back along a path and catwalks to the very base of the falls. Here Canyon Creek's waters descend through corkscrew-shaped chambers in the solid rock for more than 200 feet vertically, and then roar out through a narrow channel into the Uncompahgre River. By taking a short but steep hike, one can reach the high bridge directly above the falls. In the summer Box Canyon Park is also a favorite picnic area. In winter the chamber becomes a fairyland of ice and the nearby Uncompahgre Gorge is home to an annual ice-climbing festival.

There are at least a hundred historic buildings in Ouray. Doris Gregory's two-volume *History of Ouray* goes into great detail about these buildings, but a few will be included here. One of the most recognizable buildings is the Ouray County Historical Museum near the corner of Sixth Avenue and Fifth. The building is constructed of native stone and brick that was manufactured locally. It was originally opened as St. Joseph's Hospital on August 25, 1887, under supervision of the Catholic Sisters of Mercy. In 1896 Tom Walsh of the Camp Bird Mine helped get the hospital out of debt and on its feet again. The rear portion of the building was added in 1905. The money for the hospital was raised by donations, much of it coming from the local mines. A fee of a dollar a month would get a miner or resident treatment at the hospital in case of severe illness or accident.

The Sisters operated the hospital from 1887 until economic problems compelled them to sell it to Dr. C. V. Bates in 1920. Dr. Bates operated the facility as "Bates Hospital" until he retired in 1946. At that time the Idarado Mining Company acquired the hospital, but in 1964 the State of Colorado demanded that the hospital have extensive upgrading (one thing they required was the installation of elevators), and the mining company closed the building instead. St. Daniel's Church, which is located next-door, was fearful that the old building might be put to some detrimental use, and prevailed upon the Pueblo Diocese to purchase the building as a protective measure. In March 1976 the Pueblo Diocese sold the building to the Ouray County Historical Society for a nominal amount. The museum offers an excellent view of the area's past. Photographic displays throughout the building recall facets of local history with the surrounding mines and mining camps being shown in their heyday.

The Rev, Montgomery Hogue held the first Episcopal services in Ouray in the courthouse in 1877. Parson Hogue later held services in the local saloons with a pistol strapped around his cassock. Eventually a church building was constructed on Fifth Avenue near Fourth Street. St. John's Episcopal Church is the oldest church building

Dr. W.W. Rowan was not only the town's doctor and a leading citizen, but he owned the drug store, which also had a post office in the rear. His shelves were stocked with the latest patent medicines as well as plenty of whiskey. Courtesy Denver Public Library, Western History Dept. *(F31255)*

still standing in Ouray. In fact it is the oldest "public" building in town. It reflects the stonework of Cornish stonemasons. What was originally the church basement was built of stone, but the parishioners gave up on the planned building and roofed over the basement in February 1880. The Cornish, as well as the Welsh and English, were among the first who came to work in the mines. The carved woodwork in the sanctuary is typical of many small churches in rural England. Matching the early stonework of the church is the connecting vicarage/parish hall building, which was built in 1977.

At the corner of Fourth Avenue and Fourth Street is the First Presbyterian Church. The Presbyterians were the first to build a church in Ouray—it was only the second church built on the Western Slope of Colorado. The Reverend Sheldon Jackson and Reverend George M. Darley from Lake City established the church in the summer of 1877. The church building was dedicated October 14, 1877. Reverend Darley, a carpenter, personally directed construction upon what is now the site of the St. Daniel's Catholic Church. The original building was lost at a foreclosure sale in 1884, and the present Presbyterian Church was built on today's site in 1890. Although fire seriously damaged the interior in 1943, it was restored and many of the furnishings saved. The eastern annex was added in 1948 and another large addition was made in 1997. The bell in the church came from the 1883 Ouray schoolhouse that was torn down in 1938.

On Main Street, between Fourth and Fifth Avenues, is the Elks Club. BPOE Elks Lodge No. 492 was the first on the Western Slope, organized in 1898 with Ouray's Dr. W. W. Rowan as its first Exalted Ruler. The lodge building was built in 1904 and held its formal opening on June 6, 1905. The interior features an antique bar that came from a saloon in Ironton, and the club also contains antique slot machines (not operable), a beautiful pressed tin ceiling, an antique bowling alley (two lanes), and most of the original furnishings. The clock, by the way, always reads eleven o'clock by Elks tradition.

On the opposite side of Main Street is Wright's Hall, which was built in 1888 with cast iron piers supporting the Meskar Brothers pressed metal front. The original owners, George and Ed Wright, owned the Wheel of Fortune Mine. The second story had a large stage and a dance floor with a seating capacity of almost 500. Both local groups and renowned entertainers performed on its stage. Its seats were removable so that dances and even basketball games could be played there. Over the years locals have come to call the building "the opera house," but only a few operas were ever performed there. The second floor was remodeled in 1997 and now is used as a movie theatre. The lower floor has been used as a hardware store, a garage, and the local post office. The building to the right (north) of the Opera House was built in 1881 by the Wright brothers. It is the second oldest "public" building in Ouray. The upper story contained as many as twenty-one hotel rooms. The bottom was used as a saloon, church, and drama hall over the years. On occasion the basement was intentionally flooded in the winter to make a skating rink.

Kittie Heit built the St. Elmo Hotel, which is located directly south of the Opera House, in 1899; but first she ran the Bon Ton restaurant in what is now the patio area to the north of the hotel. Kittie Heit worked hard and eventually made enough money from her restaurant that she was able to build the St. Elmo Hotel next door. A popular diversion around 1900 was to go to the Opera House, then eat at Kittie's afterwards—or vice versa. The Bon Ton Restaurant is now housed in the basement.

The three-story Beaumont Hotel stands on Main Street at the corner of Fifth Avenue. It is perhaps the premier building in Ouray, both in appearance and in its history. The construction of the building started in 1886, but financing was a problem throughout the building period and "cost overruns" were a fact of life. The hotel opened on December 15, 1886. It contained forty-six sleeping rooms, was heated with steam, and was one of the first Ouray buildings to use electricity. The Miners and Merchant's Bank moved into the southwest corner of the ground floor, the Bank of Ouray occupied the center space and a general mercantile was on the northwest corner.

The Beaumont's furnishings and decorations were lavish. The walls were adorned with fine redwood and pine paneling, gold velour wallpaper, and an art collector's dream of fine paintings. A huge two-story ballroom and dining room is on the second floor. In the lobby were several large prints of Wm. H. Jackson photographs in beautiful antique frames. The hotel also had glass cases that contained ore samples from some of the rich mines in the area. Fine Navajo blankets decorated its second floor railing. The Beaumont was closed to the public in 1966, but has changed ownership and was completely renovated and gradually reopened to the public since 2001.

On Seventh Avenue near Second Street is the Western Hotel, which at three stories is the largest wooden structure in Ouray. It was originally called "The Monte Alta." It has

Even in 1923 the "Circle Route" stagecoach from Ouray to Red Mountain was considered a real novelty. In the background on the left and middle are Kittie Heit's St. Elmo Hotel and Bon Ton Restaurant. On the right is a portion of Wright's Opera House. Author's Collection.

had three fires over the years, but none was serious enough to destroy it. It was built in 1891 by Francis Carney and billed itself as the "miner's palace, with forty-three sleeping rooms, three toilets, one bathtub, electric lights, a saloon and game rooms." It offered its services at a cheaper rate than the Beaumont (in late 1896 the room rate was $1.25 a day plus twenty-five cents for meals). The hotel became a boardinghouse in 1916 and later was used for a small museum.

The Western Hotel was at the end of the red light district that ran down Ouray's Second Street. Although the hotel itself was legitimate it did have a back exit that a customer could take directly to the red light district without being seen. The Clipper, Bon Ton, Bird Cage, Monte Carlo, Temple of Music, Morning Star, and Gold Belt offered liquor, gambling, sex, dancing, and music. About 100 girls operated out of various establishments. Today some of the smaller "cribs" have been joined together to make larger houses.

At Fourth Street and Sixth Avenue is the Ouray County Courthouse, which was built in 1888. The basement level was also used as the city hall from 1888 to 1901. The building is constructed of local bricks and has Romanesque characteristics typified by the mansard-capped cupola. A large, old-fashioned courtroom occupies most of the second floor. It was used as a setting for the courtroom scene in the movie, "True Grit," which starred the late John Wayne. Most of the original furnishings are still in the building, while hallways leading to first-floor offices are now lined with wonderful photographs of early-day Ouray County. A smaller brick building to the rear houses the sheriff's offices.

Ouray's alligators were known far and wide. They lived near the swimming pool but were unfortunately swept away by a flood. Author's Collection.

The contractor who built the courthouse was Francis P. Carney, one of the early residents of Ouray. He was destined to one day hold the second highest office in the State of Colorado. Carney not only built the County Courthouse, but he also made the brick used in the building, since he was owner and operator of the brickyard located on the Blake Placer where the municipal swimming pool and the fishpond are now located. That site supplied most of the materials for building Ouray's earliest brick structures-the Beaumont, Wright's Hall, City Hall, the Story Block, and Manion and Beaver's Saloon (now Citizen's State Bank)-to name a few. Prior to the coming of the railroad, it would have been impossible to freight in bricks by wagon and still keep pace with the construction of many of these substantial buildings.

On Sixth Avenue between Main and Fourth Streets is Ouray City Hall. It was originally built in 1900 and was a handsome building with a facade and bell tower resembling a miniature Independence Hall. It was built on the site of the Dixon Hotel, Ouray's finest and oldest until 1892 when it burned to the ground. City Hall's second floor contained many items given by Thomas F. Walsh of the Camp Bird Mine. The donations included 6,589 books, furnishings of beautifully carved wood, a mineral collection, and stuffed animals and birds. Walsh also paid for construction of the second story, which included a clock tower and gold leaf dome. In January 1950, a disastrous fire gutted the city hall and the library. With monetary donations and volunteer labor the city hall was rebuilt to a functional state, but much of the loss, and especially the library (valued at the time at $300,000 but not insured), could never be replaced. In 1988, however, the facade was generally restored to its original elegance. The building now contains a very fine public library in addition to city offices. The adjoining Emergency Services Building was finished in 1983 and houses the county's ambulances, mine and mountain rescue trucks, and the city's fire trucks. The upper floor contains a large meeting room with kitchen facilities, smaller meeting rooms, and a large, modern kitchen.

To the north of the town is Radium Springs Park and the hot springs swimming pool. It is a wonderful place to relax after a hard day of sightseeing or a workout. The area has seen a variety of uses. Originally, Johnnie Neville's First and Last Chance Saloon and Beer Garden were located near the present-day pool. The Frances Carney Brickyard and S. P. Gutshall's Lumber Yard were also located in the area. Around the turn of the century, a baseball diamond was built, which is still used today.

When excavating clay for making brick at Carney's brickyard the holes that were created seeped full of warm water, which came from many small hot water springs that surfaced in the immediate area. Though the flow to the surface wasn't large, it was sufficient to maintain a fairly constant and warm temperature in the ponds. The only dilution with cooler water came from precipitation. The ancestors of the goldfish still living in the one remaining pond (since the building of the swimming pool in 1926) were transferred there from private ponds. When the swimming pool was built in 1926 it only had tent and log "cabanas" for changing rooms, but the first bathhouse was built in 1930.

But goldfish were not the only inhabitants of these environs. In 1921 Ed Washington brought a two-foot alligator back from Louisiana to live in the ponds and a fence was built to keep the animal from roaming. Washington sent for a mate and for some ten years the pair lived in the ponds with each alligator growing to a length of more than six feet. There are no longer alligators, but the pool area is now nicely landscaped and complimented by an attractive bathhouse built in 1988. There is an admission fee, but the one hundred fifty by two hundred eighty foot pool provides one of the country's most unique bathing experiences.

Today, Ouray's economy is mainly based on tourism. To an increasing extent, it also is becoming a home for retirees attracted to the local beauty and opportunity for recreational activities. For further information on Ouray, see *A Quick History of Ouray* by P. David Smith.

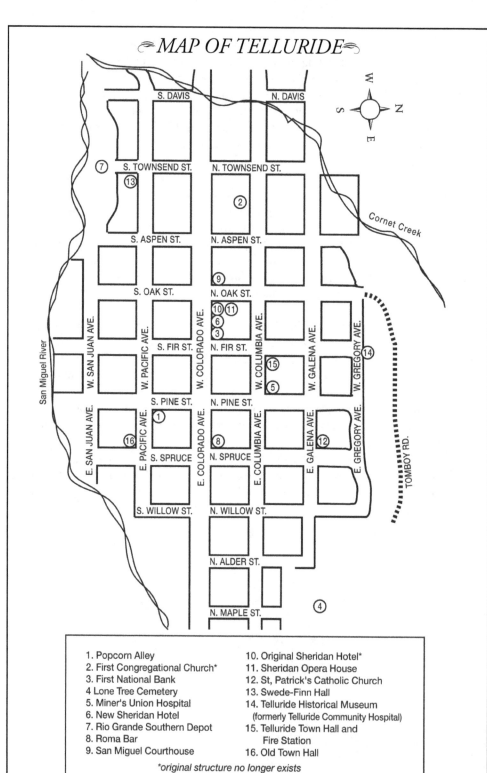

MAP OF TELLURIDE

1. Popcorn Alley
2. First Congregational Church*
3. First National Bank
4 Lone Tree Cemetery
5. Miner's Union Hospital
6. New Sheridan Hotel
7. Rio Grande Southern Depot
8. Roma Bar
9. San Miguel Courthouse
10. Original Sheridan Hotel*
11. Sheridan Opera House
12. St, Patrick's Catholic Church
13. Swede-Finn Hall
14. Telluride Historical Museum
   (formerly Telluride Community Hospital)
15. Telluride Town Hall and
   Fire Station
16. Old Town Hall

*original structure no longer exists

⤠ *Chapter Five* ⤟
THE TOWN OF TELLURIDE

The town of Telluride lies at an altitude of 8,750 feet near the headwaters of the San Miguel River near the eastern end of one of the most beautiful glaciated mountain valleys in the world. The nearby steep mountains rise more than forty-five hundred feet from the valley floor. The eye is inevitably drawn to the mountains on the east where Black Bear Road zigzags up the mountain and Ingram Falls spews forth from the mountainside. To the southwest of Ingram is Bridal Veil Falls—365 feet high. The powerhouse at the top of the falls still works, although the building is also a private residence.

Prospectors first explored the area around present-day Telluride in the 1860s, but the land was deep within Ute territory. Like most places in the San Juans, the early prospectors found signs that the Spanish had been in the area doing placer mining much earlier. Originally, the American prospectors also did considerable placer mining along the San Miguel River. The first recorded prospectors showed up in 1872 when men came over Ophir Pass from Baker's Park. They were mostly placer mining in what came to be known as "San Miguel Park," but a few men began to look for veins in the local hard rock. Lindley and Bill Remine prospected near what became Ophir in 1872 and built a cabin there in 1873. They placer mined and reported gained about fifteen dollars in gold a day from their workings. The creek that comes into the San Miguel River near the top of Keystone Hill is named after them. For several years there was no settlement in the valley, just a scattering of cabins.

In 1875 Frank Brown, Thomas Lowell, and John Mitchell started an unnamed camp on the San Miguel River. Mitchell located the Boomerang Mine, which was probably the first lode claim in San Miguel County. John Fallon discovered the fabulously rich Sheridan Mine in 1875 in Marshall Basin high above the San Miguel River. He shipped about two thousand dollars in hand-sorted ore to Alamosa that year. Almost a hundred people (including Frank Brown's wife-the first white woman in the area) lived in the beautiful San Miguel Valley by 1876.

In 1876 the lode filings started in earnest. J. B. Ingram prospected in Marshall Basin and staked the Smuggler claim between the Sheridan and Union. That same year James Carpenter and Thomas Lowthian staked the Pandora at the far eastern edge of the park. The Liberty Bell was also staked in 1876 by W. L. Cornett.

Frank Brown and John Mitchell laid out the town of San Miguel City about a mile down valley from present-day Telluride in 1876. Charles Sharman did an official survey in 1877. The San Miguel Gold Placer Company did placer mining in the nearby Mill Creek and the San Miguel River. San Miguel City was perhaps Colorado's first nature-conscious town. It was laid out with every street having a small ditch and "care was taken to not destroy the natural forest on the ground...thus the town is already provided with shade trees." Dairy cows and beef cattle grazed nearby. By 1879 San Miguel City had a population of about one hundred and by the next year the town had grown to about two hundred (but nearby Columbia was by that time larger). Although San Miguel

City started earlier than Columbia or Telluride, no plat for San Miguel City was filed until 1885.

There was considerable placer mining downstream along the San Miguel River, yet local placer mining didn't really produce much gold. Total placer mining production was about five thousand dollars a year until 1890, when it jumped up to eighteen thousand dollars to twenty thousand dollars for a few years and then was discontinued. Early lode mining wasn't doing any better. No large amount of ore was shipped in 1877-78, but what little was taken out went to Silverton and averaged 75 to 125 ounces of silver.

The town of Columbia (which would become the town of Telluride) was surveyed on eighty acres a couple of miles up the valley (east) from San Miguel City in 1878, but it was not actually incorporated until September 30, 1879. The ever-present Otto Mears was one of the original stockholders in the Columbia Townsite Company. As a result of its head start, San Miguel City was the early population leader of the two towns, but by 1880 Columbia had outgrown its neighbor. About six or seven hundred people lived in the valley by 1880, but they were only serving as a supply point for the mines higher up in the nearby basins.

A small settlement called Newport was established just below Bridal Veil Falls in 1875. After about a dozen cabins were located around the nearby Pandora Mine, the name of the settlement was changed to Pandora. The name of the company running the mine was "Folsome," so some called the few cabins by that name and the post office even used this designation for a short while. The population was estimated at eighty in 1885. Pandora had a ten-stamp mill operating by 1882, which was increased to forty stamps later. It was originally a custom mill, meaning it worked for whoever brought ore instead of handling the ore of a specific mine. However, the mill was eventually purchased by the Smuggler-Union in 1888 and connected by a tram to that mine. Eventually other mills were built at Pandora to serve the Tomboy and other local mines.

Pandora couldn't really be called a town. It was more like an industrial complex. It was a place where the mines' trams came down out of the mountains to deposit their ore, a place where the mills ground the ore into concentrates, and a place where eventually there was a railroad spur. There were barns for animals, offices for the mines, and even a few houses for the men with families and boarding houses for the other workers; but it was not really a town with stores and schools. If Pandora's residents wanted to shop, go to church, or send their kids to school they went to Telluride.

In 1879 Fossett noted that "extensive placer mining deposits are being worked on a large scale by many companies and firms. The adjoining mountains are seamed with numberless gold and silver veins." Yet there wasn't really that fifty thousand dollars in ore had been shipped out by 1880. Keystone was the biggest placer operation. It was located near junction of the San Miguel and the South Fork of the San Miguel. The Gold King Mine and several mines around Old Ophir were producing good ore and had *arrastras* working. The Gold King Mine was said to ship three tons of ore a day to Silverton at the time.

By 1880 Columbia was beginning to look like a real town. Schools had been started and two newspapers, *The Journal* and *The Examiner*, began publication in 1881. *The Journal* was originally started in San Miguel City by Ed Curry, and then moved to

This Harper's Weekly *drawing shows the area around San Miguel City about 1878(?). The artist took a little liberty to show both Ingram and Bridal Veil Falls from this angle. Author's Collection.*

Columbia. It remained in business until 1929. The local population was filled with immigrants. Typical of the San Juans, a good percentage were Cornish (or "Cousin Jacks") because the Cornish were generally well experienced in mining. There were also a good number of Italians, Irish, and Scandinavians. Columbia's first school was started in 1881. Lillian Blair taught about thirty-five children out of a rental house. Yet in 1881 the Smuggler, Mendota, Cimarron, and Argentine were the only mines shipping ore to the stamp mill at Pandora.

The town's main street (which runs east and west), named Colorado Avenue, was the center of the business district. At the far western end were some of the nicest houses. To the south of Colorado Avenue, down by the river and the railroad tracks, was the poorer part of town and the red light district. To the north of town were the churches, hospital, and school. The further uphill a person lived, the better off and more respectable the resident was considered to be.

On February 27, 1883, San Miguel County was carved out of Ouray County, and Columbia (Telluride) was named the county seat. There was great confusion at the time because the new county was originally called *Ouray* and what was left of the old Ouray County was called *Uncompahgre County.* Just four days later the name *Ouray County* moved back to Ouray, and the new county became San Miguel. The first school building was also built in Columbia in 1883 (it had only two rooms). Later, a brick building replaced the original frame. In 1883 Columbia had 400 inhabitants and was officially named the county seat.

About one hundred men were working in Marshall Basin in 1883. A small smelter at Ames ran for one year, but it didn't work well. The rest of the ore was packed to Ouray or Silverton. By 1885 Telluride had a population of 850 compared to 175 in San Miguel City. Its success was probably due in large part to being closer to the trail to the mines located in Marshall Basin. San Miguel City had been counting on the placer operations.

Pandora was a mill town. Most of the town's houses are at the lower right. Ingram Falls is at the center and Bridal Veil to the right. Author's Collection.

Ouray's Dave Day (stung by San Miguel County being carved out of Ouray County) had little respect for the new county seat. "Telluride has seven lawyers and two dance halls; 0 churches and 000 school houses. Mercy what a wicked village."

In 1885 Crofutt reported that Columbia contained "one bank, stores of all kinds, several hotels, one 20 and one 40 stamp mill, one weekly newspaper, *The News,* and a population of about 1,400." He mentioned there were ten lode claims nearby, then spent a considerable time describing the placer claims. "The placer mines along the San Miguel River and its tributaries, aggregate upwards of a hundred miles." He also pointed out the scenery:

> To those desiring to climb the most rugged mountains, visit wild, dismal and almost impenetrable canyons and "drink in" the grandest of all American Alpine scenery, and all within a day's ramble from a comfortable hotel, we would certainly recommend a visit to (Telluride).

In 1886 the new courthouse was completed, but it burned to the ground in 1887. By the end of 1887 it had been rebuilt, and it still stands today. The year 1886 also saw the construction of Columbia's first water lines. Before that time water was hauled into town using burros and barrels.

Columbia's name was officially changed to Telluride in 1887, although the post office had been calling Columbia *Telluride* since July 26, 1880. The change of name happened because the postal department was having great confusion between towns called Columbia in both Colorado and California. The abbreviated Cal. and Col. were just too hard to distinguish. The new town took its name from tellurium, which residents thought was a common gold ore in the nearby mountains but which turned out to be relatively rare. At the time of its incorporation in 1887, Telluride was already officially a "city," because the town had exceeded the necessary two thousand population it officially needed to use the title of city in Colorado.

From the time of the earliest mineral discoveries, Telluride suffered more than any the San Juan towns from the problem of lack of good transportation. Although there weren't any high mountain passes to cross (such as at Silverton), the area was located sixty miles from the nearest broad valleys or existing roads. Ore had to be packed by burros or mules over that distance because there were no wagon roads or railroads nearby. When the Utes were removed from Colorado, the D&RG Railroad quickly moved into Montrose; and the railroad had eventually been extended up to Dallas (near present-day Ridgway) and then to Ouray. But there the railroad stopped; there was no branch to Telluride. The San Miguel mines still had a fifty to sixty mile haul over the rough wagon road that Otto Mears built in 1882 to get their ore to the railroad. The Mears road was so bad that Dave Wood built his own road from Montrose in 1884. It cut three days off a typical freight trip and three hours each way off the stagecoach ride. After 1882 Telluride ore could also go over Ophir Pass to Silverton where it was then shipped out on the D&RG Railroad. However, Ophir was a trail that could be traveled only in the summer.

The history of Telluride cannot be told without telling the story of L. L. Nunn. He was a small man (five-foot one and one hundred fifteen pounds) but one of great drive and determination. He came to Telluride in 1881 and set up shop as a lawyer; but he was from out of state and didn't know much Colorado law, so his first job was shingling the roof a house. Eventually, he specialized in Colorado mining law and then gained control of the only bank in the county. He also gained an interest in several mines, purchased several commercial buildings, and built one of the finest homes in Telluride.

At the same time the nearby Gold King Mine was using staggering amounts of money to haul in coal and wood for its boilers. Nunn thought that a new invention, alternating current electricity, might be the answer. Direct current had been around for a while; in fact, it was being used by some of the mines in the Ouray area, but it couldn't be sent over long distances. Nunn talked George Westinghouse into building a small alternating current generator, which he installed at Ames. He then ran a line about two and a half miles to the Gold King Mine and immediately dropped the mines' power costs by more than eighty percent. Eventually he ran a line from Ames to Telluride and then over Imogene Pass to the Camp Bird Mine and down to Ouray. His use of alternating current helped change the world. Nunn started a school for engineers at Ames to teach them about electricity. Locally, the students were known as "pinheads." Young collegians came from all over the world to study electricity at Telluride.

One of Telluride's most notorious events occurred June 24, 1889. Butch Cassidy (George LeRoy Parker), Tom McCarthy, and Matt Warner robbed the San Miguel Valley National Bank. It was Butch Cassidy's first bank robbery. After collecting about twenty four thousand dollars they ran from the bank, jumped on their horses, and rode off toward

Since Telluride was first called Columbia, a lot of its business used the town's original name, including the proud owners of the Columbia Trading Company. Author's Collection.

Rico and Dolores. No one in the posse (which supposedly included L. L. Nunn) was especially eager to catch the riders-some of the men even told others to slow down in case they might catch "the wild bunch." Butch went to Ames and then Trout Lake where the gang divided the loot just past Lizard Head and split up. The posse gave up about Rico. The stolen money was never recovered, but Butch did leave one of his horses behind, which Sheriff J. A. Beattie rode for years afterwards "to pay for my expenses of the chase."

Telluride was not just a mining town but also a ranching town. There were a great many large ranches nearby, and on any given Saturday night it was likely that there would be just as large a number of cowboys in the Telluride saloons as miners. Stores were open until ten or eleven at night to cater to the men after they got off work.

Jim Clark, Telluride's legendary marshal who at one time was also a member of the Jesse James gang, was out of town at the time of the Butch Cassidy bank robbery. He had arrived in town in 1887 and despite his criminal background did a good job of cleaning up Telluride. He also helped many of the town's needy people, even giving them money. However, he continued his lawless ways when out of town (in fact, some people claimed he was paid off to be out of town at the time of the bank robbery). It is undisputed that he disguised himself and held up miners, tipped off outlaws to gold shipments, and committed several other criminal acts while he was the town marshall, but none of it was done in Telluride! Clark was shot in an ambush in front of the Colombo Saloon on Telluride's main street on August 6, 1895. Some of the locals believed the Telluride town council hired a hit man when Clark refused to resign as marshall. His murder was never solved.

On November 23, 1890, Otto Mears's Rio Grande Southern Railroad reached Telluride and, as in the other San Juan towns, the railroad gave great stimulus to the local mines. The population of the town and the nearby area grew to almost five thousand persons, and there were ninety businesses in town at that time. It was as if Telluride had seeds planted but no one had watered them until 1890. Then the town burst into full bloom almost overnight. There was such a boom that Telluride advertised itself as "The Town Without a Belly Ache," meaning that no one went hungry in Telluride. The final spike, which completed the Rio Grande Southern Railroad from Durango to Ridgway, was driven at Rico about a year later on December 17, 1891.

It was a proud day when the first locomotive chugged its way into the Telluride valley. The whole town turned out to welcome it. Soon the railroad hauled out an average of 150 cars of concentrates every month from the Telluride mills. The conductors came up with a slogan after the rough trip, announcing as the train pulled into town: "To-Hell-You-Ride." It was reported that the track laying around Telluride took much longer than expected as the Telluride saloons were too tempting to the RGS tracklayers.

For a long time Telluride was the latest of all the large San Juan communities to not have a church. An 1894 brochure gave a lengthy explanation as to why they had only one:

This early day Telluride photograph shows the commercial district on Colorado Avenue about 1885. The burros are loaded with supplies for a high country mine. Courtesy Denver Public Library, Western History Dept. (X149)

The fact that any given place has a multiplicity of churches does not necessitate that it is a specially religious community, very often it is quite the contrary, and conversely the statement that Telluride has only one church (the Congregational built in 1891) does not necessarily imply that we are, for that reason, irreligious, but it does show, and that to our credit, that we are not stupidly and doggedly denominational.

Gambling and saloons were popular in all the San Juan towns. This is the Cosmopolitan in Telluride. Marshall Kenneth Agnes Maclean is at the right, leaning against the bar around 1910. Courtesy Denver Public Library, Western History Dept. (X84)

In Telluride's defense, services had been held in local saloons and at the courthouse for almost a decade.

The year 1893 was a hard one throughout the San Juans. The repeal of the Sherman Silver Purchase Act meant that the U.S. government no longer bought silver for its treasury. The Rio Grande Southern Railroad was forced into bankruptcy but the Denver & Rio Grande Railroad ran the short line in receivership for many years thereafter. Passenger trains stopped in 1931, but the Galloping Goose (a contraption that was half train and half bus) ran along the route from 1931 to 1949.

The time after the Silver Panic of 1893 was not particularly hard for Telluride because many of its nearby mines also produced gold. The Tomboy Mine alone took out more than $1,250,000 between 1895 and 1896, about half of which was profit. By 1898 gold had been found in the nearby mines in such large quantities that Telluride mining became profitable again. The population of Telluride proper rebounded to 2,446 in 1900 and was up to about 3,000 persons in 1910.

By the late 1890s there were more than ninety businesses in town, and a considerable number of new farms and ranches were being established on the mesas to the west of Telluride. It was also a good time for the sporting houses. About two dozen saloons and an equal number of cribs, brothels, and "parlor houses" crowded Colorado and Pacific Avenues. The most popular was the Pick and Gad. Others included the Whitehouse, Gold Belt, Silver Bell, and the Cozy Corner. About 175 women worked at the establishments at this time. Diamond Tooth Leona and Jew Fanny were among the madams. If you think you've heard these names before, it's because most prostitutes traveled a circuit around the San Juans, moving every few years.

At the turn of the century, Telluride was a typical "Wild West" town. The saloons, dance halls, and parlor houses were open twenty-four hours a day. There were the usual gunfights and killings when men got drunk or tempers exploded, but in 1900 the newspaper bragged that no one had been lynched in Telluride yet and only two or three men had been murdered. One man tried to pull a gun on the Rico sheriff in a Telluride saloon. A bullet from the sheriff's gun cleanly blew off the man's thumb as he tried to cock his gun. Telluride Marshall Jim Knous (six-foot, five-inches tall) took care of a lot of the trouble. Generally, he would stop fights before they got started and then send the men home to cool down or sleep it off. He was so successful that he was fired for not making enough arrests, thereby lessening the fine money that helped fill the city's coffers.

Telluride also had a much more civilized side. The Telluride Silver Cornet Band was famous throughout Colorado. The twenty-piece orchestra played on the Fourth of July and at many other festivities. Dances, theater, balls, circuses, revivals, and recitals were constantly held. Sledding and ice-skating were popular in the winter. The Fourth of July was always a two or three day affair that included ball games, horse races, and mining contests as well as the usual fireworks. Another form of entertainment in the late 1890s was fire hose company races. Men pulled the hose carts for many years, because the firemen didn't have time to round up and hitch up horses. Telluride had some of the very best hose cart racers in the San Juans, if not all of Colorado.

Labor troubles between miners and the management of the mines contributed to the unrest of the period. By the turn of the century many Telluride miners were unionizing—in fact, they were leading the state in this respect. The first local union was Telluride Miners Union Number 63, which was chartered as a member of the Western Federation of Mines on July 28, 1896. On July 1, 1900, William Barney, a shift boss, deputy sheriff, and nonunion supporter, disappeared. The owners accused the miners of killing him. Later his body was purportedly found on Boomerang Hill. He had been shot eleven times and his body thrown in the brush. The union denied any involvement in the killing, but few people believed them.

The first union strike was at the Smuggler-Union Mine in May 1901. The Colorado legislature in 1899 had passed a law establishing an eight-hour workday for miners, mill workers, and smelter workers. Although the Colorado Supreme Court later declared the law unconstitutional, every mine owner in the Telluride area—except the Smuggler-Union—had accepted the eight-hour day and a three dollars per day minimum wage. The strike was mainly in response to contract mining that required the miner to buy his own supplies and the fathom system, which required a miner to make a certain amount of forward progress for his pay regardless of the type of rock or width of the vein. The miners also wanted an eight-hour workday. Nothing much was accomplished by the strike. After much posturing on both sides, work at most of the mines resumed by late July.

However, on July 3, 1901, about 250 union miners armed with rifles, shotguns and revolvers stationed themselves behind rocks and trees around the Smuggler Union property and started shooting. The nonunion forces finally surrendered with casualties amounting to three dead and six seriously wounded. The rest of the nonunion men were beat up and their shoes taken, and they were sent over Imogene Pass to Ouray with orders not to return to Telluride. By July 4 the workers at many of the mines around the state were on strike. The San Miguel County Sheriff asked the governor to send in the state militia, but instead a commission was sent to investigate. In the meantime, several buildings at the Smuggler Mine were blown up. Finally, an agreement was made. The contract system was abolished; and many, but not all, of the union demands were met. The union had won a battle but not the war; and the incident unfortunately set a precedent for how the losers of the labor battles would be treated.

Eventually, troops were sent there and also to Telluride. Restrictions such as the closing of saloons and curfews were put into effect. Local residents, as well as visitors, had to have military passes just to appear on Telluride's streets. The houses of the local population were subject to search at any time of the day or night, with or without good reason, and anyone was subject to arrest and deportation from Telluride without a trial or even a hearing.

On November 20, 1901, a load of hay, which had been stored at the mouth of the Smuggler Mine to use as feed for the horses and mules working in the mine, caught fire just as the shifts were changing. Twenty-eight union miners died of smoke inhalation. Then the Liberty Bell avalanche disaster occurred (nineteen men were killed at the mine as well as others in the nearby mountains in February 1902.) More than twenty-five hundred people were present for the Smuggler miners' funeral. The union claimed that manager Arthur Collins was negligent in not putting out the fire quicker or dynamiting the tunnel exit and instructing the miners to go out other exits. The union also noted that there were no safety doors to block the flames and smoke. The miners at

Looking east on Colorado Ave. about 1890, the flagpole and speaker's stand were very evident. Author's Collection.

the Liberty Bell blamed their management for cutting too many trees too close to the mine for mine timbers, leaving nothing to stabilize the snow. That same year a fire at the Pandora Mill claimed almost as many lives as the other disasters. Unfortunately, some mine owners let it be known that they felt the natural disasters were God's way of punishing the miners for unionizing!

One incident, which is alleged to have occurred during the strikes, bears repeating even though it may be fable. A postman supposedly went to help rescue an injured nonunion friend who was trapped at the Smuggler Union by angry union workers. The postman cut holes in his mail bag for his friends legs, put him in it, and led his friend through the angry crowd. The union men let them pass, supposedly because they were afraid to tamper with the U.S. mail.

There, a year later, Arthur Collins, manager of the Smuggler-Union, was shot dead by a shotgun blast as he sat playing cards with friends in the manager's residence at Pandora. The day before, Collins had placed an ad in the paper offering jobs to strikebreakers. No one was ever convicted of the murder. Bulkeley Wells volunteered to succeed Collins as manager. The local mine owners brought in professional gunfighters who were deputized and acted as security guards for the protection of the mines. The Telluride miners boycotted the businesses that wouldn't support them.

In his book, *Bostonians and Bullion*, Robert Livermore, mining engineer and Bulkeley Wells' brother-in-law, described Telluride as still being a "gay" town during this period.

> Telluride was filled with uniforms, miners down from the hills with well-filled pockets, cowboys up from the range their cattle shipped, their horses standing in dozens outside the town's forty odd saloons, presented a gay and festive sight. It was always a picturesque town, just a little more gala in the style of the old west than the neighboring towns of the San Juans.

In September 1903, the men at Telluride's mills struck for the same reduction of their work hours (from twelve to eight hours per day) that the local miners had received.

The next month, one hundred miners at the Tomboy walked out in sympathy when the Tomboy mill was staffed with nonunion employees. Six railroad cars of Colorado National Guard troops arrived in November, armed with a gatling gun among other weapons, and non-union workers were brought in to replace the strikers. Martial law was again declared, including an 8:00 p.m. curfew. More strikers were deported. At this point the union even considered rolling dynamite bombs into town or poisoning the town's water supply with cyanide.

The fact that mine manager Bulkeley Wells served as a captain commanding Troop A, First Squadron Cavalry of the state militia, showed the allegiance of the troops. Vigilante groups deported union workers who caused trouble. The unrest and disturbances continued for many months into 1904. It culminated in an alleged bombing attempt that did no harm except blowing Bulkeley Wells, sleeping in his bed, out of his house. The union gave up the strike on November 29, 1904. At the end of all the violence, the mine owners had won; yet the miners and mill workers won respect as well as some concessions.

Telluride was in the news for reasons other than labor unrest. In the early 1900s, Telluride boasted that it was the best-lit town in the world because it was using alternating current from the nearby Ames generating plant and also a power plant constructed on the top of Bridal Veil Falls. Because the power plant sits on top of Bridal Veil Falls, it doesn't get its energy from the falls. Instead, a pipeline extended up the mountain to a lake and delivered water under high pressure to a Pelton wheel which generated

Telluride's hose cart races were always well attended. This view also affords us a grand view of Colorado Avenue about 1895. Walker Art Studio Photo. Author's Collection.

electricity. Bridal Veil was not a commercial plant but was wholly owned by the Smuggler-Union (later Telluride Mines, Inc.) and supplied power for their mines and mill. When the Idarado Mine acquired Telluride Mines, Inc., the still-operating power plant was part of the package. Its output, however, was inadequate for all of Idarado's needs and it was closed down. It later reopened privately and today sells electricity to the local power company.

The winter of 1905-06 was extremely harsh throughout the San Juans. More than one hundred people died in the snow that winter. Some of the bodies were not found until spring. The spring runoff also caused disastrous floods. In 1909 many of the RGS bridges were washed out by a major cloudburst, and it was evident that the tracks wouldn't be repaired for months. Sheriff George Tallman happened to own the local beer bottling plant and served as the Telluride agent for Budweiser beer. He wired Gunnison that Telluride was in dire need of a railroad car of beer. On September 22, 1909, some 120 casks of Budweiser reached Placerville and were brought on to Telluride by mule train. "French Alex" led the "Budweiser relief train."

That same 1909 flood broke the Trout Lake dam. A quick telephone call to Ames warned them of the coming danger. Supposedly, the pinheads at Ames quickly calculated the estimated depth of the flood, crawled to the roof of the building they were in, and watched the water come up to within three inches of their estimate.

On July 27, 1914, cloudbursts hit the town of Telluride. Cornet Creek flooded the town with a sea of mud and debris from the huge waste dumps at the Liberty Bell Mine. The debris piled up to eight feet deep in places. One woman (Vera Blakely) was killed, the lobby of the Sheridan Hotel and the lower floors of the Miners' Union Hospital were filled with muck almost to the ceiling, and several homes were washed away. The wall of mud and muck crushed other homes. Fifty carpenters built long flumes to the river, and the power company strung lights along the streets where another hundred men used fire hoses and shovels to wash the flood debris down the flumes into the river.

In 1916 Telluride went "dry" along with the rest of Colorado, and just like the other San Juan mining towns it had its fair share of bootleggers. The city even imposed a five hundred dollar "soft drink tax" on local establishments. More than ten wholesale bootlegging operations were going at one time, and Telluride liquor was known throughout Colorado as first-class.

As in all the San Juans, the 1918 flu epidemic hit Telluride hard-almost as hard as Silverton. Ten percent of Telluride's population died in November, and then a second flu epidemic hit in 1919.

World War I was the beginning of the end for mining near Telluride because war efforts required that operations be directed away from gold and silver to base metals. Inflationary pressures also affected the price of just about every commodity except gold and silver; the price of gold was fixed by the federal government at $20.67 per ounce. Mines started closing throughout the area. The last big local mine to close was the Smuggler-Union in 1928. Although the Ajax Slide had run several times without doing damage, in the spring of 1928 it hit two of the homes located on the hillside above the Smuggler-Union Mill in Pandora, killing two employees' wives and a small child.

Pro-management employees of the Smuggler-Union Mine carry superintendent Charles M. Baker's casket in 1902. Courtesy Denver Public Library, Western History Dept. (X62726)

One of Telluride's larger-than-life legends concerns a swindle perpetuated during the crash of 1929. Bank examiners were ready to close down the First National Bank of Telluride, but President Charles D. Waggoner pulled a five hundred thousand dollar fraud on six New York banks. He went to New York, got the money, and sent enough back to Telluride to cover his depositors. Supposedly, he also kept a large amount of the money for himself. The ruse was short-lived. At the trial, his attorney characterized him as a western Robin Hood who stole from the Eastern rich to protect the savings of his poor Telluride depositors. In spite of this novel defense, Waggoner was convicted; it came out in the trial that Waggoner liked to play the stock market with his depositors' money, and some of his "investments" were nothing short of gambling. Nevertheless his depositors didn't lose a single cent, and Waggoner was a local hero—but he was still sentenced to fifteen years in prison. He was paroled after just three years, but never returned to Telluride. Despite his efforts the Bank of Telluride folded in 1934. His story is well told in *Telluride: From Pick to Powder* by Richard and Suzanne Fetter.

As bad as the times were, Telluride refused to die! The town's population dwindled to 512 in 1930, but it was blessed with a goodly supply of enterprising people who found a way to survive, keep their city alive, and provide income for many citizens. Many men in Telluride and throughout the San Juans made a legal living during the Great Depression by working small mines in the area or even panning in the river. By cutting costs to the bone and not using expensive equipment, they could make enough money to make ends meet.

Nevertheless, in 1953 the Telluride Mines announced it was closing its operations, releasing 230 workers, or ninety percent of Telluride's work force. The *Telluride Tribune* wrote, "There was a lack of foresight among us not to make Telluride a great tourist attraction to buffer such an economic blow." But the Idarado Mine bought out the Telluride

Mines a few months later and work continued. Telluride in the 1950s and 1960s was referred to as "a time capsule"—a place for the "purist" time traveler.

At about that time, mining began a revival (relying mainly on base metals) that would, in large part, bring Telluride back to good economic health. It was the Idarado Mine and Mill that kept Telluride alive during the 1950s, 1960s, and 1970s. Its "Gray Mill" processed eighteen hundred tons of ore a day. For many of those years the Idarado was the first or second leading producer of precious metals in the state. Many of the Idarado's miners lived in company houses in Pandora and worked at the large mill there. The gold and silver that were found were just added bonuses. By the 1960s Telluride's population was about six hundred—most of whom were connected in some way with the Idarado.

By the 1960s jeeping was becoming popular, and Telluride was being discovered by tourists. Imogene, Ophir, and Black Bear Passes were all reopened as jeep trails, but that wasn't enough. A new enterprise again came along just in time to "save" the town. This time it was a recreation industry based on snow and fantastic scenery. Telluride had tried ski runs built by volunteer workers, but in 1968 Joseph Zoline of Beverly Hills, California, started the Telluride Ski Area, which opened up in late 1972. The ski runs used a lot of names from the past—Smuggler, Tomboy, Pandora, and Pick and Gad to name a few.

Telluride has fast become a world-class cultural resort. In the summer months, among many other things, the city hosts a festival almost every week. The Telluride Film Festival was first held at the Sheridan Opera House in 1974 and has grown into an international event. The Telluride Bluegrass Festival began in 1973 and is now attended by as many as twenty-five thousand fans. Telluride has been touted as the "new Aspen," an international resort and home to the rich and famous. But one thing remains unchanged. The town is surrounded by some of the most beautiful mountain scenery in the world.

It is worthwhile to take a historic tour of the town. The three-story brick Sheridan Hotel, at the corner of Colorado and Oak, is one of the most prominent of the Main Street commercial buildings. It was built in 1890, and the third story addition was added in 1899. The Sheridan Opera House was added in 1914. It still has its beautiful old balcony and box seats, and some of the original stage backdrops are also still present. The seats could be slid under the stage to open the place up for dances or other gatherings. The Opera House seats about two hundred and is connected to the hotel by a second floor walkway.

The Sheridan has been beautifully remodeled over the years. In its time the hotel rivaled the Brown Palace in Denver, serving gourmet foods and rare wines. The cherry wood bar was imported from Austria, and the walls of the bar are covered with fine calfskin. In the Continental Room diners could dine in one of sixteen curtained booths, calling for their waiters by telephone. William Jennings Bryan, on July 4, 1903, gave his "Cross of Gold" speech outside the hotel. He ran for office on the platform of reinstating the Sherman Act. "You shall not press down upon the brow of labor this crown of thorns; you shall not crucify mankind on a cross of gold." Bryan won overwhelmingly in Colorado but unfortunately lost the election. Lillian Russell and Sarah Bernhardt also stayed at the hotel.

Robert Livermore in *Bostonians and Bullion* described the Sheridan in the early 1900s.

(It was) the mecca of visiting travelers and the elite of mining men and cattle-men, who liked a slightly better brand of liquor or cigars at the bar.... Usually a well-patronized poker game went on at the Sheridan, between cowmen with well-filled pockets and the local talent. One such, I remember, lasted twenty hours, with lunch and other refreshments served at the table.

Telluride's present town hall at Fir and Columbia was originally the town's school, built in 1883 and later serving as a recreation hall. The first school in Telluride had been in a house and attracted fifty-three students in its first year. In 1895 the existing eight-room school replaced the original two-room structure. The fire station was directly behind city hall. The tower was used for stretching out and drying the long fire hoses.

Telluride's first church, the Congregational Church, with the high steeple and stained glass windows, was dedicated on February 15, 1891. Its first minister was the Reverend N. S. Bradley. Like all the San Juan towns the first services were held in the Telluride's saloons. The first church service in Telluride was held by Parson Hogue in Jim Hurley's Corner Saloon. (Hogue was also the minister at the Episcopal Church in Ouray).

The wood-framed St. Patrick's Catholic Church at Galena and Spruce was built in 1896 and is still in use. It was the second church building built in town. By 1899 it had two hundred members. It is at the corner of East Galena Avenue and North Spruce Street. The area around the church became known as Catholic Hill because of the large number of Catholic Italians and Austrians that lived there in the early 1900s. The wooden figures of the Stations of the Cross were carved in Austria.

The brick San Miguel County Courthouse at Oak and Colorado was originally built in 1886 but burned to the ground and was rebuilt the next year. It was used for dances, literary meetings, and a variety of other community uses besides court and the county clerk's office. Before the courthouse, the county offices were on the second floor of a local saloon!

One of the original Galloping Geese is displayed outside the courthouse. It was donated to the city when the RGS closed in 1952. These half-car and half-locomotive contraptions economically carried six to seven tons of freight and half a dozen passengers and kept the Rio Grande Southern Railroad operating for many years past its prime.

The red sandstone hospital was built by Dr. Hall at the upper

Charles (Buck) Waggoner is on the left in this photo, which was taken to show the bank's stability. Courtesy Denver Public Library, Western History Dept. (X101)

Telluride's main street (Colorado Ave.) looked a lot different about 1940 than it does today. Courtesy Denver Public Library, Western History Dept. (X63116)

end of Fir Street in 1893 and served as such until 1964. It was well known as the Old Miner's Hospital. Harriet Backus, author of *Tomboy Bride*, gave birth to her first child in this hospital. It later became a dormitory for skiers. After extensive renovation and restoration it has recently reopened and is now the San Miguel County Historical Museum.

The town's red light district, known as "Popcorn Alley," was on East Pacific. Three of the original cribs can still be seen. Because men outnumbered women at least five to one in most Colorado mining camps, prostitution was bound to exist. The city condoned it and, in fact, licensed it. Fees generated by saloons and bordellos were a major source of revenue. The Senate was a popular saloon/bordello run by "Big Billy"—a very popular (and large) madam. It closed in 1935. Jack Dempsey worked at the Senate as a young man. The Silver Bell was a popular dance hall on East Pacific. The Pick and Gad was a "big house"—a classier version of the bordello where the women dressed fancier, as if that mattered! Other establishments were the Big Sweede, Idle House, Cozy Corner, and the White House. Most of the red light district's buildings and many of the town's saloons burned in a large fire.

The Rio Grande Southern depot (down by the river on Townsend) was constructed in 1891, shortly after the railroad arrived, on land donated by the city. It was abandoned in 1952, but was restored in 1991 and is now used as a restaurant. The restoration even included putting a new foundation under the old building.

The three-story brick building at the corner of North Pine and Columbia Avenue was built in 1902 as the Miner's Union Hospital. It was built in large part to offset the political influence of the mine owners' hospital. The hospital lasted only two years, but the building was used later by Railway Express, the Elks Club, and, in 1922, the post office.

The second building on the north side of Colorado between Fir and Pine was originally the site of the San Miguel Valley Bank—the bank Butch Cassidy robbed. The original building burned. The building a couple of lots to the east with the Ionic columns and capitals was the Bank of Telluride, where Charles Waggoner pulled his 1929 swindle.

The stone building at Fir Street and Colorado once had a tower on top but it was weakened severely by a 1914 flood and was removed for safety reasons. It used to house the First National Bank. The sandstone for the building came from a quarry up Cornet Creek.

This Muriel Wolle photograph from circa 1940 shows the grandeur of the First Congregational Church, even though she was getting a little old. Courtesy Denver Public Library. (X83)

The Swede-Finn Hall still stands in what was called Finn Town at Townsend and West Pacific Avenue. Dances, parties and balls were held for Telluride's large Swedish and Finnish populations. Finn Hall, a competitor, was located only two lots to the east. There were several hundred residents of each nationality in the town at one time, and they loved to get together to dance, eat, and socialize. Finn Town Flats, originally a boardinghouse, was on the south side of Pacific Avenue.

The Roma Bar is at the corner of Spruce and Colorado. It is one of Telluride's oldest bars and contains an 1860 carved walnut bar with twelve-foot mirrors in the back bar. It was completely renovated in 1983.

The old stone town jail at Pacific and Spruce was built in 1885 and was just recently vacated by the public library. The original wooden jail was also built on this spot in 1878, but it is now located at the Town Park.

The town's cemetery is at the northeast end of town. Originally, the cemetery was in an aspen grove high above the valley floor because everyone expected the lower areas to be destroyed by placer mining. George Andrus donated the land for the present cemetery when his two year-old child died. The oldest grave dates back to 1878.

MAP OF YANKEE BOY & IMOGENE ROADS

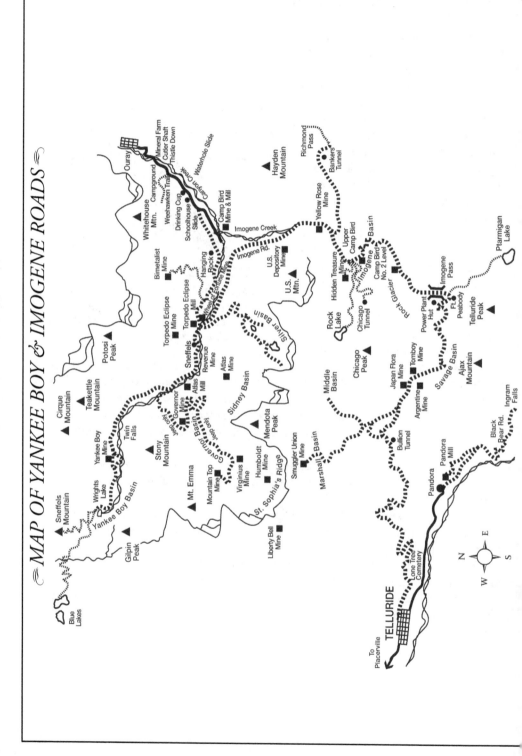

Chapter Six

IMOGENE, YANKEE BOY AND TOMBOY

An all-weather road (Colorado 361) closely follows Otto Mear's toll route southwest out of Ouray to the famous Camp Bird Mine and then fragments into several four-wheel-drive roads that lead into some of the most beautiful basins in the San Juan Mountains. Because of the rugged terrain, the road that leads over Imogene Pass to Telluride is the only route that is not a dead-end. Because this part of the road tops out at over thirteen thousand feet (it's the second highest pass in Colorado), Imogene Pass might be open only from late July to early September. The eighteen-mile trip from Ouray to Telluride takes about three hours of steady driving, but with the side trips it's an all day affair.

In the early 1900s, at the zenith of local gold mining activity, the Camp Bird Road was constantly lined with John Ashenfelter's wagon teams of six or eight huge draft horses—their drivers sometimes strapped to their seats to keep from falling out. Some days Ashenfelter had as many as two hundred to three hundred burros or mules and a hundred wagons on the road, most of them carrying ore and supplies between the Camp Bird or Revenue Mines and the City of Ouray. On the way down, the back wheels of the loaded ore wagons might be fitted with "rough-locks"—a device in which the rear wheels could be cradled and skidded along the road. On the underside of the rough-lock were sharpened "dogs" or claws that could bite into the roadbed. On slanted roads the rough lock might be applied to the high wheel only, which had the effect of making the wheel skid upgrade or at least keep it going straight. During the first decade of the twentieth century Ashenfelter not only brought down ore but also delivered as much as three million pounds of supplies and a hundred thousand board feet of lumber to the Revenue and Camp Bird Mines each year.

Today, instead of burros and wagons, there is a steady stream of four-wheel drive vehicles filled with tourists bound for the cool mountain meadows, a midsummer sea of brilliant wildflowers, and hundreds of cascading waterfalls of pure mountain snow. Unfortunately, the route has become too popular. In an effort to slow down the number of visitors, the Forest Service has instituted a fee that must be paid by all who enter the area. Opponents of the fee point out that the Forest Service provides only minimal services and that public land should be open to all citizens without a fee unless corresponding services are provided. Proponents argue that the fee helps keep the area from being overrun and destroyed by tourists. Public officials are lining up on both sides of the issue, so there might or might not be a fee station at the time you read this book.

After leaving the town of Ouray on Highway 550 head south and after rounding the first hairpin curve, depart to the right (south). Keep to the left through the highway department grounds and cross the new and higher bridge across the Uncompahgre Gorge. The first (lower) bridge leads to Box Canyon Park. The entire gorge is an icy wonderland in the winter when locals turned it into Ouray's Ice Park by taking water and spraying water down the steep sides. Ice climbers come from all over to hone their skills on some of the best ice climbs in the world.

About a half a mile from the bridge, just beyond the city water tank, are several private roads leading left (southeast). The first two originally provided access to the Mineral Farm Mine—one of Ouray's earliest mining operations, having been discovered in the fall of 1875. Outcroppings of ore-bearing rock appeared haphazardly all over the sloping terrain and could be "harvested" by digging trenches-hence the name "Mineral Farm." August W. Begole and John Eckles sold the mine in October 1878 for seventy five thousand dollars. The *Ouray Times* reported excitedly in its December 14, 1878, issue:

> The transfer makes an epic in the history of mining in this section and ushers in the dawn of prosperity for our town and people. Business will undoubtedly be better and money more plentiful this winter than ever before in the history of this town, and we may congratulate ourselves that a company of such magnitude and all so business-like in its management, has cast its lot in our community.

A variety of minerals were mined, including ankerite, barite, chalcopyrite, copper, and galena. Fossett in 1879 declared the Mineral Farm to be "one of the wonders of this state." The mine operated into the 1930s at which time it employed thirty to forty men under the management of William "Bill" Cutler. Total production was valued at over one million dollars. In recent times the mine was used in the movie "True Grit." The snake pit scene, as well as several others, took place at the Mineral Farm. The property today is not accessible to the public.

About a mile and three quarters from town a jeep road turns sharply to the left (near the entrance to the Chalet Hayden Subdivision) and ascends for several miles through

The Mineral Farm Mine did a lot of its mining using tranches to follow the vein. This photo shows the mill complex. Photo courtesy of Denver Public Library Western History Dept. (F3287)

heavy aspen groves to the head of the Hayden Trail. At the end of the public portion of the road, and slightly beyond the trailhead, is the head frame of the Cutler Shaft, which was originally connected by a now-collapsed tunnel to the Camp Bird Road. William Cutler operated the mine briefly in the 1930s in connection with the Mineral Farm. The Hayden Trail is a very steep but spectacular foot trail that leads over Hayden Mountain and then descends into Ironton Park. There is less elevation to conquer if you travel from the Ironton Park side to the Camp Bird Road, but it is a long hike and portions of the trail are not easy to follow. Plan on the hike taking all day.

Back on the Camp Bird Road and directly north from the Canyon Creek Bridge is a small campground and a short hiking trail that leads about a half mile back into the Angel Creek area (originally called "Quaking Asp Flats.") Most of this area is public property and camping is allowed at the Angel Creek Camping Pod; but it is also a favorite spot for bears, so be careful. At the time of this writing it is also the location of one of the controversial fee stations.

Continuing up the Camp Bird Road, several gulches and avalanche areas are visible to the south across the creek. The first is Squaw Gulch, so named because the spire rising to the right of the gulch looks like a squaw carrying a papoose. The second gulch contains Lewis Creek, and Thistledown Creek flows down the third. At this point a power pole can be spotted at the very top of Hayden Mountain. The pole is what remains from a turn-of-the-century power line when electricity was brought to Ouray and its surrounding mines from the newly built alternating current generators at Ames near Telluride. The Thistledown Mine is visible about half way up the gulch. M. L. Thistle discovered the mine in the early 1900s, and it was well financed and included a tram, mill, and boardinghouse. It even had its own hydroelectric plant, but its ore proved to be low-grade. It was reopened after World War I to produce fluorspar, which is a soapy, greenish to purple mineral used as a flux in metallurgy and in making hydrofluoric acid, opal glass, and in some enamels. Until recently, the Thistledown boardinghouse stood at the site of the present-day campground. There are also toilet facilities at this site.

The Weehawken Trail leads up the edge of the creek on the other side of the road until it splits. The right fork leads to the Alpine Mine with wonderful views of Ouray, and the left fork leads into beautiful, isolated Weehawken Basin at the foot of Potosi Peak and Whitehouse Mountain. Either hike could be made in a half day.

After crossing Weehawken Creek the Camp Bird Road goes through a series of switchbacks and then passes along a high shelf road, where it traverses cliffs that are hundreds of feet high. This part of the road stymied the county commissioners before they turned construction over to Otto Mears. At one point a small stream gushes from a rock above the road. In the past the rock projected above the road, and pure spring water fell to the road. When the large freight wagons of John Ashenfelter plied the road, the thirsty freighters would stop for a cool drink. Many left their drinking cups on spikes hammered into the rock, so the place took the name "The Drinking Cup." Today it is remembered much more for the steep drop-off on the other side of the road. A pull-out makes it possible to see all the way down the canyon, if you are brave enough to drive out that far!

After the shelf road several private roads branch off to the left, and then the road crosses Senator Creek with its beautiful little waterfall. Then after the switchback and

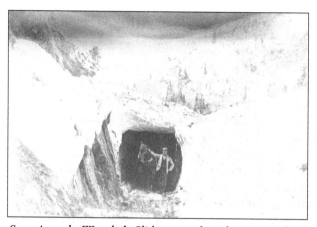

Sometimes the Waterhole Slide ran so deep that a tunnel was needed. Photo by Johnston. Author's Collection.

another shelf road, the road and the creek are almost at the same level. This point is known as the Waterhole Slide. It takes its name from the troughs that were kept at this spot to allow the freighters to give their teams a drink of water after ascending the steeps grades. In the winter, avalanches often block the road at this point. Snow can fill the canyon to such an enormous depth (up to fifty feet) that snow, ice, and avalanche debris can often be seen all summer. Many men and animals have lost their lives at the Waterhole or the Schoolhouse Slide (which is just a little farther along the road).

Three-tenths of a mile above the Waterhole Slide (and about five miles from Ouray) the road forks. The right fork (Ouray County Road 26) goes to the ghost town of Sneffles, Yankee Boy Basin, and through Imogene Basin to Telluride. The road to the left leads to the Camp Bird Mine and Mill. The mill was recently moved to, of all places, Mongolia, and the tailings pond is being reclaimed. About all that is left is the portal of the main haulage tunnel and a few Victorian buildings, which were once the homes of the mill and mine superintendents.

Between 1896 and 1902 the Camp Bird Mine produced over four million dollars for Tom Walsh. Then he sold it for another six million dollars! For many years it was the most profitable mine in the state of Colorado. Walsh was soon fabulously rich, and the mine became one of the most famous in the world. Within a few months of discovering the mine, Walsh built a large mill in Imogene Basin; but he found it too difficult to ship ore in the winter, so he later built the mill here at the confluence of Sneffles and Imogene Creeks, two miles below the mine. For many years an aerial tramway connected the mine in Imogene Basis and the mill on Canyon Creek. There was a small town located at the lower mill site with offices, storerooms, houses for management and their families, boardinghouses, assay offices, a dairy, a store, post office, hotel, recreation hall, and many other facilities.

Helen Downer Croft in *The Downs, The Rockies and Desert Gold* points out that the Camp Bird Mill often extracted more than ninety percent of the gold from its ore (the normal recovery rate at the time was only about sixty percent to seventy percent) because:

(The ore) was known to be one of the docile for treatment. The Camp Bird Mill had sixty stamps and the weight of 850 pounds dropped 6 to 8 inches one hundred times a minute, crushing 180 to 190 tons of ore daily. The pulp, after screening, was discharged upon silver-plated copper tables fifty-four inches wide and sixteen feet long, upon which quicksilver had been spread. This held the particles of gold.... Twice a day the amalgamated gold was recovered by scraping the greasy slime from the rectangular plates. This was carried to a retort, which was closed, then a valve controlled by turning a wheel released into it steam, the heat of which carried off the mercury. Beautiful gold ingots, 900 fine, remained. These bars were cooled, were packed into an iron chest and taken under armed guard in a coach. After one armed robbery, instead of two guards, four rode horseback with rifles at ready, and there were no further violent scenes. From Ouray the ore was sent through an express company to the mint in Denver.

The tailings went to a cyanide plant for further refining by holding the mineral-laden slime in vats holding 250 tons each. Another two to four ounces of gold as well as eleven to fifteen ounces of silver were extracted. The concentrates were then sent to Denver for treatment. Over the years the mine collected $1,660,000 from the waste and between the mill and the cyanide plant made a total recovery of ninety-three percent to ninety-five percent of the mine's gold and silver. The use of cyanide was an important advance that most of the San Juan mines picked up on. Because of the increased recovery ratio, relatively small amounts of gold in low-grade ore were worth milling. In addition, the demand for base metals such as copper, lead and zinc were expanded dramatically as the United States progressed as an industrialized nation and electricity came into use.

By 1897 the Camp Bird was using electric power that was brought in from Telluride. The seventeen miles covered by the power line was at the time the longest in the United States. The large Victorian houses at the mill site were built after an English syndicate, called "The Camp Bird Limited," bought the mine from Walsh in 1902. Between 1902

The Camp Bird Mill is in the center of this photo. The cyanide mill is to the left and living quarters are at the bottom right. The mines tunnel is in the upper section, left of center. Author's Collection.

and 1916 the syndicate made another twenty three million dollars from the mine, with a profit of fifteen million dollars.

After selling the mine Walsh moved to Denver and then Washington, D.C. His daughter Evalyn married Ned McLean, whose father owned the *Washington Post*. The newlyweds bought the Hope Diamond while on their honeymoon, and after living the life of the rich and famous, the "curse" of the diamond eventually struck. One of their children died, and Evalyn ultimately found herself penniless. Her life in Ouray, her father's discovery, and the story of her fanciful life is told in her book *Father Struck It Rich*.

The superintendent's house was begun in the summer of 1902. It had hot water heat and electricity. Most of the ground floor rooms contain twelve- or fourteen-foot ceilings. The house had chandeliers and even a fire-fighting system. A smaller house was built for the mill superintendent, and three even smaller houses for other employees. The boardinghouses built for the miners had hardwood floors, hot and cold running water, steam heat, first-aid stations, fire protection, marble washbasins, a library, flush toilets, and billiard rooms. Walsh believed that his men should be well treated!

The lower Camp Bird area was subject to avalanches. In 1906 two avalanches ran above the mill to the southeast (one off Hayden Mountain and one off U.S. Mountain), joined together, and came much farther than ever before. The avalanche "broke the mill building like an eggshell" and damaged a good part of the reading room at the boarding house. Luckily, only a few men were working in the mill at the time. One man was killed and one injured, but six more men were killed during a massive and deadly subsequent fire.

The Camp Bird continued production until 1916, when mining was halted to drive an eleven thousand-foot tunnel. The tunnel allowed access to the upper workings in Imogene Basin at a lower level and without going outdoors. The tunnel also allowed the mine to drain its water by gravity instead of using expensive pumps. The mine was worked off and on until the late 1970s, but it now appears to have been worked out-although some old miners swear there is still plenty of gold left in the Camp Bird.

Back at the fork before the Camp Bird, the right-hand road begins a steep climb into high mountain basins of unbelievable beauty. Drivers should be particularly alert for other vehicles on the narrow shelf road that passes under Hanging Rock—obviously named for the cliffs that hang out over the entire right-of-way, often dripping cold water onto unsuspecting visitors. The next few miles include glimpses of many mine ruins across the canyon, the Highland Chief being one of the biggest. A ladder coming down the canyon side was rumored to be have been built by a miner who used it as a shortcut to get to his sweetheart at the Camp Bird.

If you look carefully to the right, a rough road travels to the Bimetalist Mine, which was located in November of 1896 by Malcom Downer (who was an accountant at the Camp Bird Mine and who prospected in his spare time). The Bimetalist vein (so named because it contained good amounts of both gold and silver) was narrow, but it ran as high as one hundred ounces of gold and one thousand ounces of silver per ton. In just five years the Bimetalist Mine made Downer a wealthy man.

Shortly after the Bimetalist and a little more than six miles from Ouray, a road dips to the left and crosses Sneffles Creek. This is the main road over Imogene Pass to

The Camp Bird had its own school. The children are sitting on kindling brought to the school to help light the school's stove. Author's Collection.

Porter's Store was small but carried a variety of necessary and luxury items. The left hand corner in this photo was the post office. Author's Collection.

Telluride, and I will return to this part of the trip later. To the right is Potosi Peak that rises to 13,790 feet. Slightly up the road on the right are the remains of the Torpedo Eclipse Mine. What is now seen as an abandoned mill was once the commissary, but in the 1950s equipment was installed for milling Torpedo Eclipse and Ruby Trust ore. Before that time the ore was sent by a tram across the small valley to the huge Revenue Mill.

The Wheel of Fortune Mine (located south across the creek) was one of the first mines discovered in Ouray County. W. H. Brookover, Mason Greenleaf, and George and Edward Wright made the extremely rich discovery on October 7, 1875. It had an eighteen- to twenty-inch vein of up to twelve hundred ounces of silver and twenty ounces of gold. Its ore averaged eight ounces of gold and 175 ounces of silver. The mine was sold in 1877 to B. J. Smith and A. J. Hoyt for $160,000, but then work ceased because of disagreements between its owners. The Revenue Mine eventually worked the Wheel of Fortune from underground.

The small flat area just a short distance farther was the site of the town of Sneffles. It was also called Porters or Mount Sneffles at earlier times. Early settler George Porter was a photographer, storeowner, and postmaster of the small settlement. Sneffles was named for the tallest mountain in the area—Mount Sneffles, which in turn was named by the Hayden Survey party in 1875. Yet the mountain cannot be seen from the site of the ghost town. Sneffles was a spelling variation of the mountain *Snaefel* in Jules Verne's book *A Journey to the Center of the Earth*. Andy Richardson and David Quinn built the first cabin at Sneffles in 1875. They had come over from Silverton and discovered enough rich ore that they spent the winter.

The post office opened at Sneffles on October 31, 1879, and continued until April 3, 1895. Mail was delivered six days a week. The settlement included several stores besides Porter's and about fifteen residences. The 1885 population of Sneffles was about one hundred. All that remains of Sneffles today are several modern buildings from the Revenue Tunnel and a few older structures. Only the foundation remains of the huge Revenue Mill, but the mine still contains large amounts of valuable ore waiting shipment until the time when silver prices rise to an acceptable level.

The Revenue Tunnel came about because of the rich Virginius Mine, which was started far above Sneffles at an elevation of 12,500 feet in Humboldt Basin. The mine was next to impossible to reach in the winter and was located in an area of dangerous snowslides. Eventually it was not feasible to pump water from its deep shaft, and there was no wood available so far above timberline to power its boilers. So the Revenue Tunnel was bored from Sneffles for a distance of seventy-eight hundred feet. It was started in 1888 and completed in 1893 and intersected the Virginius vein two thousand feet below the surface. The new tunnel also allowed mining of the Monarch and several other rich veins.

The prime mover of the Revenue-Virginius was Albert E. Reynolds, who in 1880 had joined forces with John Maugham and the Thatcher brothers (who were bankers from Pueblo) to purchase the Virginius and begin large-scale operations. The old adage that it takes money to make money is never truer than in mining. Mining engineer Hubert Reed was hired to take charge of the management of the Virginius. The ore got richer but because of high costs it was 1884 before a profit was made. Then to help decrease

The Sneffles commercial area is at the left. Ashenfelter's barns on the right and the Revenue Mill in the middle. Author's Collection.

costs the Reynold's group shifted to electricity produced by a Pelton water wheel (it paid for itself in a year). Otto Mears brought a good wagon road up Canyon Creek, and a big concentrating mill was built and began producing in 1888. The concentration raised the amount of silver per ton from thirty-five to 181 ounces. The arrival of the railroad at Ouray further dropped transportation costs.

The Revenue Tunnel was the final cost-saving factor. When they hit the Virginius' vein in 1893 the mine paid profits of at least $330,000 a year until the end of the century. Reynolds and Reed became famous in mining circles. (See Lee Scamehorn's *Albert Eugene Reynolds* for the full scope of Reynold's efforts.)

The population of Sneffles increased from one hundred in 1890 to 442 in 1900. The population remained at 144 in 1920 and twenty-one in 1930. After that time the widespread use of the automobile allowed workers to live in Ouray and commute to the mine. The Revenue Mine was worked as recently as the late 1980s, and altogether the Revenue has produced more than twenty-eight million dollars in ore—mostly silver, not gold!

The huge sixty-stamp Revenue Mill was built in 1894. The Revenue was the first Colorado mine to light its interior and power its cars by electricity (direct not alternating current). Electricity was so new that the mine's boarding house had no switches and early as 1905 the mine used electric cars underground—the bare wire carried eight hundred to nine hundred volts, which resulted in the death of several miners who accidentally touched it.

The administrative personnel of the Revenue Mine and their families lived in a few houses scattered around town, while the rank and file filled three four-story boarding-houses. Each of the boardinghouses held about 100 men. They contained steam heat, electric lights, baths, barbershops, a library, and reading rooms. A miner's pay at the Revenue was $3.00 to $3.50 a day. Sixty cents a day was taken out for room and board. The local school had twelve to fifteen children in 1898.

The Revenue was the scene of one of the San Juans most gruesome mining disasters on December 18, 1896. Five Swedish miners were near the top of the Virginius shaft and were sent down in a bucket—however, someone forgot to connect the bucket to the cable, and they fell 1,100 feet to their deaths. The bodies were terribly crushed and jammed together beyond recognition.

Less than a half mile beyond the Revenue and on the other side of Sneffles Creek is the remains of the Atlas Mill. The mine operated into the 1930s, and a large three-story boardinghouse stood in the flats until about 1950. At that time it was torn down and the lumber taken to Ouray to build a motel. The cable from the mine is still connected to the mill. The mine produced gold and silver from the 1890s to the 1930s. The mill was unique for its time, and many mining journals wrote of its processes. Today, there is a Forest Service campground at this site.

Just a few tenths of a mile beyond the Atlas, the road forks again. Both roads eventually dead-end and are definitely for four-wheel-drive vehicles only. The right road is fairly easy for a jeep road and leads to Yankee Boy Basin, which was formed by glaciers that only receded about ten thousand years ago. It is only a little more than a half-mile to the high meadows and lush carpets of alpine wildflowers that make the visitors feel that they have truly reached paradise. Few areas on earth have a more spectacular backdrop for the short but spectacular summer show of hundreds of varieties of colorful blossoms, dominated by thousands of the state flower—the columbine. Yankee Boy also contains Twin Falls, a familiar scene in many Colorado photograph books. Please do not pick the flowers that bloom from late June through August, and please keep your vehicle on the roads. A tenth of a mile or so past the Twin Falls are restroom facilities.

The remains of the Yankee Boy Mine are slightly more than a half-mile above Twin Falls. As you enter the upper part of the basin, take the right fork (the left just loops out and ties back in to the main road a short distance later). William Weston (a famous British metallurgist and graduate of the Royal School of Mines in London) and George Barber discovered the Yankee Boy Mine in 1877. It reportedly produced fifty thousand dollars its first year, a great sum in those days. Fossett exclaimed, "In 1878 the lowest mill run gave 235 ounces of silver per ton, and the highest 1,700 ounces, the rich ore being from six to eight inches wide." In 1879 it yielded fifty-six thousand dollars to its lessee, F. B. Beaudry, which was the record for a San Juan mine that year. Twenty tons of Yankee Boy ore averaged 1,231 ounces of silver. William Weston sent out glowing reports of the area's mineral potential to mining journals all over the world. The publicity was invaluable for the San Juans. By 1881 the Yankee Boy Mine had eight hundred feet of tunnels and was sold to a New York syndicate. It produced gray copper, ruby silver, and brittle silver and sorted ore from the mine averaged 103 to 396 ounces of silver per ton. The mine was worked until shortly after the turn of the century.

Another two tenths of a mile farther the road forks again. The left road leads a short distance into the basin and dead-ends. The right fork is steep and rocky-be extremely cautious if you're driving. The road dead-ends at a little over a mile at Wright's Lake and the Blue Lakes Trailhead, from which one can climb Mount Sneffles. William Wright was a Cornish miner who did assessment work on the mine near the lake for many years. After his death his son Arthur kept up the work until about 1950, but no great amount

Stony Mountain rises to the right above the Atlas Mill. The boarding house is at the lower center of the photo. Courtesy Denver Public Library, Western History Dept. Photo by McClure.
(MCC2211)

The Yankee Boy Mine gave the spectacular basin its name. The mine's rich ore helped establish the San Juans as a top mining area. Courtesy Denver Public Library, Western History Dept.
(F12720)

The Ruby Trust Mine lies slightly below Yankee Boy Basin and has been very subject to avalanches over the years. Courtesy Denver Public Library, Western History Dept. (X61970)

of valuable ore was found. The Blue Lake Trail leads over the thirteen thousand-foot saddle that connects Mount Sneffles with the lesser-known Kismet. The beautiful Blue Lakes are on the other side of the divide. The trail to the top of Sneffles begins to the right (north) just before the Blue Lakes Trail begins its steep climb up and over the ridge. Be sure not to turn too soon. It seems like every year someone makes this mistake and ends up on Teakettle or Gilpin Mountain, which are traversable by only the most experienced technical climber. The trail up Sneffles is not difficult (although a steady climb for about two hours), but if in doubt seek directions.

Returning back several miles to the left fork of the Camp Bird Road (just past the Atlas Mine) Ouray County Road 23A leads quickly to the Ruby Trust Mine, which has been worked off and on right up to present. In the 1890s and 1900s the Ruby Trust produced some very high-grade ruby silver (an ore that appears to "bleed" when broken open.) The mine was close enough to Sneffles that many of its workers roomed there; three men died at the site in 1886 when an avalanche hit the mine's boardinghouse. In 1903 another avalanche destroyed the bunkhouse and badly damaged the mill. The men in the bunkhouse escaped. Their pet cat was trapped for thirty-two days but lived! Later, an avalanche destroyed a small mill set up by Emil Leonardi in 1948.

Traveling across the creek, the road leads up and around the base of 12,689-foot Stony Mountain (a barren volcanic plug) and up Mendota Peak to Governor Basin. A remote weather station is encountered after about half a mile. It reports the weather by satellite; most of the information is used to predict avalanches. A short distance farther, the gated road to the right leads to Governor's Basin and the base station for the Mountain Top Mine, which closed its main operation in 1929 but is still being worked off and on today. Near this gate was the location of the Governor Mine. A tram used to run into Humboldt Basin to the Mountain Top Mine, which is located at twelve

The Virginius helped prove an axiom that the higher up a vein was discovered the more likely it was to be a very rich find. Courtesy Denver Public Library, Western History Dept. (F4023)

A rescue crew poses with an avalanche victim in Governor's Basin about 1885. Note their skis planted in the snow. Courtesy Colorado Historical Society. Photo by Vaughan Jones. (CHSX7896)

thousand feet at the northwest corner of the basin high above. Because of snowslide danger, a sixty-ton-a-day mill was constructed nine hundred feet underground in the mine. It was said to be the first such operation in the world, but it failed to operate properly. However, an underground mill such as the Humboldt's might be the way of the future. If metal prices were to rise high enough, mining would certainly start up again in the San Juans, yet environmentalists are properly worried about damage to the environment. An underground mill, which dumped its tailings back into the mine itself, could well be the answer.

Proceeding upward around the Mendota Peak switchbacks, you might notice a rough road (not really meant for vehicular travel) that leads into Sidney Basin (originally called "Son-of-a-Bitch Basin.") No mines of any consequence were discovered here other than the Atlas, but it is a beautiful spot. Traveling upward to the west we reach the point where the road starts down again into the basin. The Virginius Mine dumps are visible high on the mountain to our left (reachable only on foot), at about road level in the same gulch is the Terrible Mine, and the Humboldt Mine is on the steep slopes just under St. Sophia Ridge straight ahead (southwest). The road leads down to the Mountain Top, which is visible below and across the basin to the west. A foot trail used to lead from the Humboldt over the ridge to Marshall Basin near Telluride. Father J. J. Gibbons in his book *In the San Juans* describes in horrifying detail the perils for a man brave enough to try this route—but it saved many miles so many tried it. The jagged pinnacles of St. Sophia Ridge rise to over thirteen thousand feet behind the Humboldt and the Virginius and are visible from Telluride. The trail is now unsafe to climb for all but the experienced climbers because of crumbling rock.

William Freeland located the Virginius on June 28, 1876. It was sold and resold over the next few years at ever-higher amounts. It very quickly became one of the San Juans biggest producers. A dozen men worked two shafts and three levels in 1876, and fifteen men worked through the winter of 1877, living in cabins that were sometimes completely buried in the snow. The mine had its own post office that claimed to be the highest in the world, and a small store was also located at the site. All supplies, including wood and coal, had to be brought in during the summer. The mine produced at least thirty thousand to fifty thousand dollars every year since it was discovered until A. E. Reynolds bought it in 1880. At that time thirteen men were working the mine. The mine was rich—it averaged eight ounces of gold and 175 ounces of silver in its early years, but it got even richer as you went deeper. By 1880 the Virginius ore averaged $385 per ton, but its ore needed to be packed to the Windham Smelter north of Ouray or, for a better recovery rate, shipped all the way to Denver. Despite the danger, more than a hundred miners lived in a three-story boardinghouse at the upper level of the mine. In 1883 a slide hit the boardinghouse, doing considerable damage and killing four men.

There were other dangers too. Rev. J. J. Gibbons in his book In the San Juans tells the tragic story of Billy Maher. Billy constructed his cabin about a half mile from the Virginius at a place where tundra was the only vegetation and marmots, chipmunks, and ptarmigan were the only wildlife. During the winter his wife stayed in town. One cold winter day Billy was thawing eight sticks of dynamite on his stove. As crazy as it sounds, the act was necessary because the explosives wouldn't detonate if they were too cold. This time the powder exploded. Billy's right hand was blown off, his clothes were ripped off, and Billy was left deaf and blind. Billy's partner took off immediately to get help at the Terrible Mine. Billy's dog went uphill to the Humboldt Mine but couldn't get any of the miners to realize that he wanted them to follow him. It took Billy's partner eight hours to get to the Terrible because of the deep snow. Four men immediately left to go to Billy's cabin and found him still alive but on the verge of death. They built a sled and set off through the darkness, but a new storm had blown in. They made it to the Terrible where they spent the rest of the night. The next morning they telephoned (yes, they had phones even at this remote location in the 1890s) the Virginius to see why four of their men had not come as promised. Only then did they find out that the men had left! A rescue party found all four dead in an avalanche. One man, Allen McIntyre, had lived for five or six hours before he froze to death, as shown by about a foot of melted snow in front of his face. The other three had died instantly. None was more than four feet under the snow, and one of the men was merely inches from the surface, but the snow in an avalanche sets up like concrete. Despite the men's heroic and deadly rescue efforts, Billy Maher died of his injuries.

The Terrible was just down slope from the Virginius. It carried good silver values and had several buildings around its tunnel. Its first full year of production (1886) it made $18,300 profit, and it was later incorporated into the Virginius.

The Humboldt Mine at one time employed 180 men. Shortly before World War I the best of its claims were sold to the Smuggler Union Mine, whose main operation was located on the Telluride side of St. Sophia Ridge. The purchase came about when the owners of the Humboldt realized the Smuggler had been mining Humboldt ore from beneath its workings. The Smuggler was given the choice to pay the appraised value for the ore or to buy the claims. They chose the later. The Humboldt was another mine plagued with snowslides, and in 1898 its foreman tried to travel in a snowstorm that was so thick that he walked to his death over the side of a five hundred-foot cliff.

Leaving this remote area and traveling way back down to the ghost town of Sneffles, you will come to the turnoff for the Imogene Pass Road. Its upper stretches are usually one of the last jeep trails to open during the summer (it is usually passable only from late

Miners often traveled to the upper Camp Bird on the ore buckets but they didn't usually hold hands. Actually the left bucket is going up and the right is going down full of ore. Marvin and Ruth Gregory Collection. Author's Collection.

This is only one shift of the men working at the Upper Camp Bird Mine in 1904. Marvin and Ruth Gregory Collection. Author's Collection.

July through early September) because it is the second highest vehicular pass in North America. Only 13,180-foot Mosquito Pass near Leadville is higher. So find out whether the pass is open before you try to get to Telluride. The road crosses the creek (take the bridge if the water is high) and starts through a heavily wooded area. The immediate fork to the right leads to Silver Lake Basin, which has some really beautiful scenery, including a rock glacier and a turquoise-colored lake. Be careful because the road is a little rough. Back on the Imogene Road, the mining debris is part of the Hidden Treasure Mine. Soon the road becomes bumpy and muddy, passes through several small streams, and then edges along the top of the cliffs overlooking the Camp Bird Mine with a beautiful view back down the valley toward Ouray.

Just past where the road comes out of the trees (a little over a mile after leaving Sneffles) is the dump of the U.S. Depository Mill. The mine of the same name is located about halfway up the mountain to the right. The mine was discovered by George Wright in 1877, but it was abandoned because of constant snowslides. However, Wright made enough money to become very prominent, and he built Wright's Hall in Ouray. The mill was later used by the Camp Bird Mine off and on over the years.

Beyond U.S. Mountain, the road crosses Imogene Creek. The scenic buildings to the right belonged to the Yellow Rose Mine, which eventually became an important part of the Camp Bird. One of its tunnels opens right at road level along the shelf road that leads to the Upper Camp Bird. Shortly before the road again crosses the creek, a road to the left leads into Richmond Basin. The Bankers National Mine is located near the head of this basin. This mine did well until most of the structures were destroyed by avalanches in 1905. A faint trail near the end of the road can be followed over the top of the ridge and down into Ironton Park.

Back at the Imogene Road, cross the creek (or use the bridge in high water) and shortly thereafter emerge into Imogene Basin. The settlement of Weston was started in 1877 in this area. William Weston and George Barber discovered the Una and Gretrude claims that same year. Weston prospected in the area for only a few years, but his cabin showed up on the map for many years thereafter. Hubbard and Caleb Reed were offered half ownership of the Una if they developed the Gertrude. The Reed brothers found good amounts of silver and were given the Una, yet none of the parties recognized the true value of the mine.

Andy Richardson had entered beautiful Imogene Basin from Silverton in 1875 and named it after his wife, Imogene Jiliff Richardson. He must have loved the basin as much as his wife because Richardson occupied the basin year round from the time of its discovery until 1949. Twenty years after he first entered Imogene Basin, Richardson helped Tom Walsh with the discovery of the Camp Bird Mine.

The Hidden Treasure Mine, which is located on the west side of the basin, was one of the first large operations in the area. W. H. Brookover and Edward Wright (George's brother) came from Silverton and made their way through early season snows by snowshoes to stake their claim on October 7, 1875. The first hand-picked ore ran as high as twelve hundred ounces of silver to the ton! In 1880 Fossett reported, "The ore bodies are usually large and permanent, with galena the base of the metals. The lower workings are said to show better mineral than was found on the surface." It was one of the most prominent veins of the entire San Juans and was worked into the 1930s. The Walsh family held on to it when they sold the Camp Bird, and they still own it today. Snowslides and winds have reduced all the mine buildings to piles of rubble. Higher up in the basin behind the Camp Bird is the Chicago Tunnel and higher still is Rock Lake.

Walsh's story of the discovery of the Camp Bird Mine began in 1896 when he was looking for ore to use as flux in the small smelter in Silverton of which he was a part owner. Although Walsh had already had considerable success in mining, he had been hit hard by the Silver Panic of 1893, and he was close to bankruptcy. His prospecting trip eventually led him into Imogene Basin, where he noticed Andy Richardson puttering around one of the old dumps. With Andy's help Walsh inspected most of the basin. Later, when Walsh was recovering from an illness, he asked Andy to break through a snow bank to explore one of the old tunnels and get him some samples. Walsh later rode up to the mine, examined several sacks of ore that Andy had taken from the vein, and went in and got his own samples alongside a shiny vein that Andy had been sampling. Walsh suspected that the less showy minerals might contain gold-and they did to the tune of three thousand dollars per ton. Walsh bought the Gertrude and Una Mines for ten thousand dollars and managed to buy almost every claim in the basin before his secret got out. Walsh gave his collection of claims the name "Camp Bird" from the jays that stole part of the Walsh's lunch when they picnicked on the spot.

The Camp Bird Mine itself started at the Number 3 Level, which is located to the southwest of the mill dump. Walsh started by reworking the dumps of the Gertrude and Una Mines. This gave him the money to increase his development. Shops, warehouses, and a three-story boardinghouse were built in Imogene Basin, which Tom Walsh furnished in a very fine manner for the day. The boardinghouse had a recreation room and library, and the bill of fare at the dining room included all kinds of food besides the usual meat and potatoes. There was even a post office from April 1898 to March 1918. The original postmaster was none other than Andy Richardson. Walsh paid his miners well—$4.50 a day at a time when the going rate was $3.50. Eventually the Camp Bird produced over fifty million dollars in ore. A tradition at the Camp Bird was that every traveler through Imogene Basin was offered an excellent meal at no charge; but Imogene Basin could get cold, and the snow piled deep in the winter. David Lavender in *One Man's West* does a wonderful job of describing a miner's life in this high basin.

These women examined Ft. Peabody after it was abandoned. The round cairn held the flag in stiff winds and the men held out in the small structure to the left. Courtesy Denver Public Library, Western History Dept. (X60245)

Camp Bird miners loved to tell tales of men whistling in the basin in the winter and their notes freezing. When all the notes thawed out in the

The snow piled up deep at the Tomboy in winter. In fact the Tomboy holds the record for the most snow in one year in the San Juans (581 inches). Author's Collection.

spring, the woods sounded like a steam calliope, the story went. The snow became so deep that the packers placed rags in the tops of trees to blaze trails. If a horse went off the trail it might collapse in the deep snow and have to be dug out or sometimes even killed if the packers couldn't dig it out. In the summer the miners told visitors that the flags at the tops of the trees were put there by the birds so they could find there way home on a cold, snowy night.

Only a horse trail went farther up from the Camp Bird to Imogene Pass, and then to Tomboy, until 1965 when the jeep trail was established. Avalanches posed a constant problem in Imogene Basin. The double chevron avalanche defense system that was built of rock above the Number 2 Level is still visible. Above that is the Number 1 Level (the Una) and the 0 Level (the Gertrude). Snowslides still did their damage. In April 1900 one man was killed and one was injured by a slide. Again in March of 1902, a slide killed one man and injured four others. Then on February 24, 1936, Rose Israel (the cook), Chap Woods (the mill superintendent), and Ralph Klinger (the camp blacksmith) were all killed when three or four slides ran at once. The mill, bunkhouse, and other buildings were all severely damaged.

Proceeding up the divide, the road slants across steep talus slopes. There is a great viewpoint about half way up. Then the road traverses muddy hills as it reaches the final approach to the pass. The old power lines and poles, visible on the ground, originally brought newly discovered alternating current all the way from Ames. From near the top of the pass Ptarmigan Lake is visible to the south as well as the three Red Mountains. The dead-end road (usually gated) down to the lake is rough. The building at the lake was built for the men who kept watch on the pumps that brought water over the mountain to Savage Basin and the Tomboy Mine. The pumps were powered with electricity. The Highline Trail (hiking only) leads from the lake down to the Red Mountain area or another branch can be taken to the Black Bear Road.

Visible at the very top of the pass is Fort Peabody. It is little more than a pile of stones surrounding a shack that supported a flagpole from which a flag was reported to still be flying when the new jeep road over Imogene was opened in 1968. The fort was built during the labor troubles in 1904. For a short while a squad of state militiamen was stationed at Fort Peabody to prevent deported union "troublemakers" from returning to Telluride over the pass after they were shipped out to Ridgway and then walked to Ouray. The fort got its name from Governor Peabody, who sent the troops to Telluride in response to the labor problems.

The top of the pass (13,114 feet) gives commanding views of Savage Basin, the Sneffles Range to the north, and the San Juans in general. One can see all the way to the La Sal Mountains in Utah. The shack on the ridge was for the electrical linemen who were sometimes stationed in the corrugated hut during the winter, so they would be ready if any problems occurred with the transmission line. Electric heaters heated the shack.

The road going down to Telluride is generally much easier than the road on the Ouray side. Savage Basin (named for Charles Savage, who prospected the area in 1876) is a glacial cirque in which the road levels out at about 11,500 feet. The concrete flumes were recently mandated by the Environmental Protection Agency in an effort to reduce pollution. Scattered everywhere in the basin are the ruins of what is left of the mining camp of Tomboy, named for Otis C."Tomboy" Thomas, who located the Tomboy Mine in 1886. The mine was hardly worked until 1894, but after that time the valley was filled with mine buildings, a large sixty stamp mill, livery stables, a school, homes, machine shops, a store, a three-story boardinghouse capable of handling 250 men, and even a bowling alley and tennis courts.

The Tomboy began producing valuable ore in 1894 and was sold to the Tomboy Gold Mining Company for a hundred thousand dollars that year. Its vein was four to twelve feet wide, three thousand feet deep, and one thousand feet long. Later, a tram was built that ran eighteen hundred feet down from the mine to the mill. The pipeline brought water from Ptarmigan Lake over the ridge and down to the mill. Later, a tunnel was driven in from the mill and replaced the tram. The Tomboy later bought the Argentine, which was also very profitable. In fact, it was the best and the biggest vein in Tomboy Basin! When the Argentine vein began to run out in 1911, the Tomboy bought the Montana and then the Sydney-White Cloud Group in 1927.

The property was sold to the Rothchilds of London for two million dollars in 1897. In 1902 its large sixty stamp mill was built at the east end of the basin. From 1898 to 1915 the Tomboy paid an average dividend to its stockholders of 15.1 percent. In 1902 mining engineer F. M. Downer pointed out, "The Tomboy...should be perhaps classified first in this district as a dividend producer but does not rank as high in point of tonnage as either the Liberty Bell or Smuggler-Union." Its vein was eight to sixteen feet wide but averaged only fifteen to sixteen dollars per ton-almost all gold.

The Japan was also located in Tomboy Basin and had its own mill. It became prominent in 1897 when it shipped $112,500 in silver, $85,000 in gold, and $50,000 in lead. The next year it increased its production to $312,500. In later years it was called the Japan-Flora. This mine was on an extension of the Tomboy vein (northwest and 2,000 feet above).

At one time the Tomboy had about thirty houses for married workers. Winter life in Savage Basin is vividly described in Harriet Backus's *Tomboy Bride* and Duane Smith's *A Visit With the Tomboy Bride*. It was a never-ending battle against cold, deep snow and avalanches. Backus reported snowdrifts that built up to twenty feet. The mules that went to Telluride with concentrates strapped to their backs delivered her supplies on the return trip. Poor Mrs. Backus even found her silverware rearranged on the floor by packrats.

There was a lot of time for winter recreation in Tomboy on skis and snowshoes. The miners even came up with the idea of putting wings on their sleds in an attempt to fly, but their "hang gliders" never worked well. The Tomboy sponsored dances and brought up "respectable" women from Telluride for the event.

Mel Griffiths, in *San Juan Country*, tells a fascinating tale of how the ore was shipped from the Tomboy to Telluride at a time when there were robberies in the area. Miners often rode horses up to the mines and let the animals return to the stable on their own. The Tomboy put their gold in the horse's saddlebags and then unloaded the ore at the other end. No one knew the riderless horses were carrying a fortune.

The Tomboy eventually constructed a tram to Pandora where its concentrates were shipped out on the Rio Grande Southern Railroad. The mill at the southeastern end of the basin was the Columbia-Menona (Iona). The neat rows of concrete pedestals that you can see today were for the Tomboy boardinghouse. To the west of the boardinghouse was the Tomboy Mill.

The Tomboy closed in 1927, most of its buildings were razed, and much of its machinery was sold for scrap during World War II. Work was resumed underground (but not on the surface) when the Tomboy and several nearby mines were consolidated into the Idarado Mine in the early 1950s.

From the lower end of Savage Basin there are great views of the Telluride Ski Area, the Wilsons, Lone Cone, and other views all the way into Utah. As you leave Savage Basin a road (usually locked) leads to Marshall Basin to the northwest. The basin's most famous mine was the Smuggler, which was so named because it was located in an improperly surveyed area between the Sheridan and the Union claims (they each staked about five hundred feet too much land in 1875). J. B Ingram staked the Smuggler in 1876. He didn't work the Smuggler but leased it to John Donnellan and William Everett, who also located the Mendota Mine above the Sheridan. John Fallon had staked the Sheridan on August 23, 1875. He also leased it to John Donnellan and William Everettt. Their first shipment of ore, packed out by burros, is said to have brought ten thousand dollars. Fallon then sold the mine for fifty thousand dollars to people who in turn sold it to an English syndicate from Shanghai, China. Within just a few years the Sheridan, Smuggler, and the Union Mines were combined. A little later all of the claims in Marshall Basin were added to the group. Hand-picked Smuggler Union ore ran eight hundred ounces of silver and eighteen of gold per ton. Its high grade ore averaged $225 per ton, and its concentrates ran $425 per ton. In the 1890s tunnels were bored at lower levels, and the Sheridan Incline (track laid on the ground) went down from Marshall Basin to Pandora. The incline dropped 2,200 feet in its 6,900-foot length. The cars were tied to cables and could carry three hundred to four hundred tons of ore a day. The present-day jeep road passes through about the middle of the incline, which gives you some idea of just how high it went!

Because of the early discoveries Marshall Basin was the site of most of Telluride's early mining activity. The Marshall Basin vein was four to six feet wide, three thousand feet deep, and four miles long. In order of elevation the Union was the lowest mine, then the Smuggler, then the Sheridan, and the Mendota was at the top. However, there was still the problem of transportation. Before the arrival of the RGS in Telluride only the richest ore could be shipped (some going to Ouray and some going to Silverton).

By 1895 the lower Bullion Tunnel entrance was the main way that Marshall Basin ore was shipped. The incline ran into problems from that point down in the deep snows, so a 6,700-foot tram was built. In 1899 Robert Livermore took over as manager and

The Smuggler-Union's buildings are visible in the center of the photo with the Tomboy being in the background. Courtesy Denver Public Library, Western History Dept. (X62312)

Once Smuggler-Union ore started coming out of the Bullion Tunnel in the 1890s, a collection of mine buildings were scattered across the hillside. Courtesy Denver Public Library, Western History Dept. (X62314)

the workforce was expanded to three hundred (even as many as nine hundred men at times). A boardinghouse, offices, post office, and general store were built at the Bullion Tunnel site next to the tramhouse. Later, a crusher house, stables, and even a hospital were built. All of this was near the place in the road with the big flat spot—about a half-mile down from Tomboy. In 1907 the Pennsylvania Tunnel was built seven hundred feet lower that the Bullion Tunnel. Another tram connected to the mill. The buildings at this site completely burned in 1927, were rebuilt, but burned again in 1975. The site of the Pennsylvania Tunnel can still be seen far below the road near the bottom of the gulch. The Cimarron Mill and Mine were also near the Bullion Tunnel, but just a little below it (and the present road). The owners of the Revenue Mine on the Ouray side of the divide ran the Cimarron. Its mill had forty stamps.

The population of the little settlement of Smuggler was 110 in 1900 and 315 in 1910. The mine eventually became one of the largest in Colorado with over thirty-five miles of underground workings and a total production of over twenty-two million dollars. One of its gold veins was over four miles long and was mined to a mile deep. Although the mine produced millions in extremely rich ore, it made no profits since there were huge operating expenses. The Smuggler Union ceased operations under its own management on December 1, 1928 after fifty-five years of operation, but it was later incorporated into the Idarado Mine and worked underground for many more years. Savage and Marshall Basin mines, along with the Liberty Bell Mine, accounted for ninety percent of the San Miguel County mineral production over the years, and their veins were so large and

The Liberty Bell was well established about 1900 when this photo was taken. Courtesy Denver Public Library, Western History Dept. (X62359)

rich that San Miguel County ranked third in statewide production. Only Leadville and Cripple Creek did better.

Telluride had massive labor problems centered around the Smuggler-Union. As detailed earlier in Chapter 5, the difficulty started in 1901 when union miners struck for an eight-hour day and the abolition of a pay system that was based on how much ore was removed. The two sides battled for years.

The present-day road from the Tomboy to Telluride was built in 1902. The Tomboy Road was considered one of the greatest engineering feats of the time. Before its construction the trail zigzagged down the steep gulch to Pandora, where the mills were. Trams did most of the hauling of the ore.

Not far from the Smuggler, but about two miles down the road from Tomboy and high up in the next basin, is the Liberty Bell Mine, which was discovered in 1876 by W. L. Cornett. The Liberty Bell is on Cornet Creek, named for (but misspelled) for the discoverer of the mine. St. Sophia Ridge is behind the mine, and the rich Virginius Mine was on the other side of the ridge. The Liberty Bell was not greatly productive (and at times was considered almost worthless) until 1897 when a rich vein was discovered. A tram was then built to the mine's forty-stamp mill which was at the east end of Telluride in the valley below. The mine produced a lot of gold and had a large cyanide mill built in 1900 at the site of the stamp mill. It ceased operations in 1921 with sixteen million dollars in ore to its credit. The Idarado took over the Liberty Bell in 1953.

A massive avalanche struck the Liberty Bell on February 28, 1902. The slide carried away the boardinghouse, the ore house, and the tram station. Nine men were killed and eight were injured. When the rescue team was recovering the bodies, a second avalanche hit, killing six more men and injuring one. Then on the way to Telluride the rescue party was struck by a third slide which this time killed four more men and injured five. A total of nineteen were killed and eleven injured. Some of the bodies were not recovered until spring. There could have been many more deaths, however, because the slide barely missed the bunkhouse where thirty to forty men were sleeping.

Throughout the years many deaths occurred as a result of snowslides at the Liberty Bell or Smuggler-Union and throughout the San Juans. In fact, there have been so many deaths due to snowslides in the San Juans that only the most disastrous can be reported here. (John Jenkins' book *Colorado Avalanche Disasters* tells many of these stories in detail.) To add insult to injury, lightning struck the Liberty Bell Mine's ore car rails that summer and killed three men deep inside the mine.

The Tomboy Road (Forest Road 869) continues to switch back about four miles down the mountain, passing through the famous "Social Tunnel" (also called "Marshall Tunnel") about a mile below the Smuggler-Union's Bullion Tunnel, crossing several small streams and opening up spectacular views of Bridal Veil Falls, the switchbacks of Black Bear Road, and Ingram Falls on the east headwall of the valley. The road is relatively smooth but narrow and steep, and there are many places where two cars could not pass. Wonderful views of Pandora (to the east of Telluride), the Telluride Ski Area, and the Mountain Village can also be seen before the road finally ends in the middle of the Telluride residential district on the north (uphill) end of Oak Street.

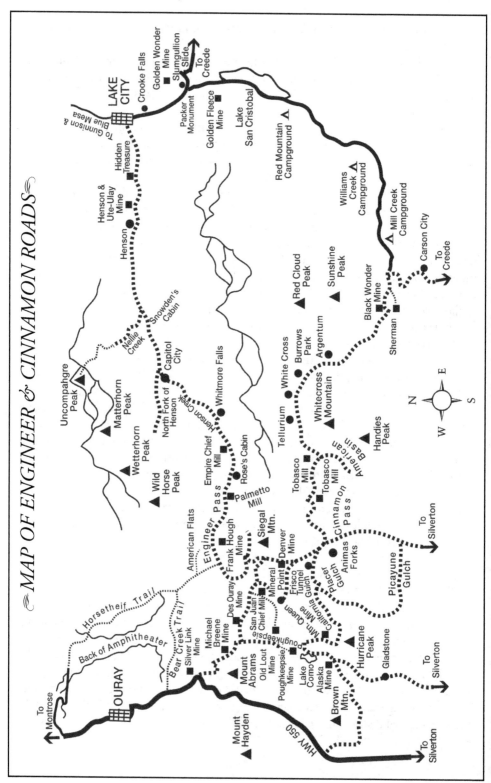

MAP OF ENGINEER & CINNAMON ROADS

LAKE CITY

OURAY

Crooke Falls
Golden Wonder Mine
Slumgullion Slide
To Creede
Packer Monument
Golden Fleece Mine
Lake San Cristobal
Red Mountain Campground
Williams Creek Campground
Mill Creek Campground
Carson City
To Creede

To Gunnison & Blue Mesa
Hidden Treasure
Henson & Ute-Ulay Mine
Henson
Snowden's Cabin
Nellie Creek
Capitol City
North Fork of Henson
Henson Creek
Whitmore Falls
Red Cloud Peak
Sunshine Peak
Black Wonder Mine
Sherman

Uncompahgre Peak
Matterhorn Peak
Wetterhorn Peak
Wild Horse Peak
Empire Chief Mill
Rose's Cabin
Palmetto Mill
Tellurium
White Cross
Burrows Park
Argentum
Whitecross Mountain
American Basin
Handies Peak
N
W   E
S

American Flats
Engineer Pass
Frank Hough Mine
Siegal Mtn.
Tobasco Mill
Tobasco Mill
Cinnamon Pass
To Silverton

Horsethief Trail
Back of Amphitheater
Bear Creek Trail
Silver Link Mine
Michael Breene Mine
Des Ouray Mine
San Juan Chief Mills
Mineral Point
Denver Mine
Frisco Tunnel
California Gulch
Mtn. Queen Mine
Placer Gulch
Animas Forks
Picayune Gulch
To Silverton

To Montrose
Mount Hayden
Mount Abrams
Old Lout Mine
Poughkeepsie Mine
Lake Como
Alaska Mine
Brown Mtn.
Hurricane Peak
Gladstone
To Silverton

HWY 550
To Silverton

126

⇌ Chapter Seven ⇌
ENGINEER AND CINNAMON PASSES

The Cinnamon Pass Road from Lake City to Silverton was built in 1874, although the top portion was really just a pack trail at that time. In fact it was bad enough that two years later in 1876 the *San Juan Prospector* declared, "There is no wagon road between Lake City and Silverton." On August 10, 1877, the first stagecoach made it over Engineer Pass, and that route then became the accepted way of travel between the two cities, unless a traveler was on horseback or on foot. The trail (later the Henson Creek and Uncompahgre Toll Road) from Ouray up to Mineral Point was one of the last roads to be built into the high mining areas. Without a trail or a road it took early prospectors two summers to work their way from Mineral Point down Uncompahgre Creek and Bear Creek into the bowl that came to eventually hold the City of Ouray.

Today, the western access across the Engineer Road begins at 8,850-foot elevation at the spot where the Uncompahgre River crosses under U.S. Highway 550. The trip from Ouray to Lake City requires about four hours and the return via Cinnamon Pass to Silverton takes about an equal amount of time. However you might make this a two-day round trip because you will be tempted to stop often along the way. The road is four-wheel-drive only, and in some of the sections in Ouray County are very rough. It is also steep-rising four thousand feet in ten miles. However, the trip is always worth the trouble, because the view from the top of Engineer Pass provides a spectacular panorama of the San Juan Range, and there are many abandoned mines and ghost towns along the way. There are also some good primitive camping spots, but your stay is limited to a maximum of fourteen days in a thirty-day period.

Starting from the Million Dollar Highway, the road (Ouray County 18) climbs rapidly through a series of switchbacks as it follows the Uncompahgre River to the southeast. This road was the original route into Ouray, and the creek was considered a continuation of the Uncompahgre River even though bigger Red Mountain Creek follows U.S. 550 south toward Silverton. The Uncompahgre has in headwaters at Lake Como in Poughkeepsie Gulch. The first two miles of the Engineer Road contain some of the roughest terrain on the entire route. If you can make it through this section you should have no trouble with the balance of the trip.

After traveling about a half mile, look up to the left to see what remains of the Silver Link Mine, which was located high above the canyon at 10,500 feet. Little more than a few dumps is left today. It looks as if it would be impossible to get to the mine, but it was extensively developed in the early 1890s and produced an estimated hundred thousand dollars in silver and copper before the year 1902. More than twenty-two hundred feet of drifts (tunnels) followed the Silver Link vein that was sometimes up to twenty feet wide. Hand-picked ore from the mine contained up to thirty percent copper and three hundred ounces of silver per ton. A little less than a mile past the switchbacks, there is a gated road (private property) that leads back to the Silver Link.

The main road then passes along a steep shelf (one of the scary parts of the trip) almost three hundred feet above the river below-straight down! Be careful—it

The Silver Link Mine was located in one of the steepest, most inaccessible places in the San Juans. Author's Collection.

is impossible for two vehicles to pass throughout most of this section. An interesting low-grade vein appears in the cliffs about halfway through the shelf road. If you ever wondered what a mineral vein looks like up close, then this is your chance to see one.

Just after you pass by the small waterfall in the crevice to the north of the road, you will be going through the workings of the Mountain Monarch and the Mickey Breene Mines (originally called the Michael Breene). These claims were located on September 20, 1874, by Milton W. Cline, one of the founders and first mayor of the town of Ouray. He sold his interest to Frederick Pitkin (later governor of Colorado) and William Sherman, and the mine was sometimes called the Sherman-Pitkin Mine. The Reed brothers managed the mine and did development work from 1887 to 1889. It wasn't until 1890 that a mill, boardinghouse, and power plant were constructed at the Mickey Breene. Silver and copper are the chief metals in its veins. Hand-picked ore averaged 658 ounces of silver per ton. The building that is left standing today is actually the ore-loading bin of the Mountain Monarch, although the area is still generally known as the Mickey Breene. The mine was worked in a big way during the 1890s, with David Reed acting as its chief engineer and general manager. Even in recent years, and after many starts and stops, there has still been some production at the mine.

About a half mile past the Mickey Breene, Diamond Creek runs across the road. Immediately thereafter are several small campsites in the flats below the road, but no facilities are available. The Fish and Game Department sometimes stocks the stream around the campsites with trout, but your luck will probably depend on a recent stocking.

About a mile past the Mickey Breene, the road into Poughkeepsie Gulch (originally called Alpha Gulch) forks to the south, and the main road to Engineer Pass curves to the northeast over bedrock. The area around the fork was once officially laid out as the town site of Poughkeepsie, and there were a few cabins in the vicinity. There are several good primitive camping sites along the creek but no facilities. Although the road up Poughkeepsie can be traversed for a most of its distance, it is rough and should be regarded as a dead end because of the extremely dangerous terrain at the upper end.

When the town of Ouray was first founded, the Poughkeepsie Gulch road was the shortest and easiest way to get to Silverton-except it was so rugged and steep that it could be traveled only on horseback or by foot. After passing over Hurricane Pass at the top of

The Mickey Breene Mine's complex included the large boarding house to the left and its mill to the right. Author's Collection.

Poughkeepsie, the travelers followed Cement Creek to Silverton. Wagons or the stage had to travel to Mineral Point and then down the Animas River to Silverton. Poughkeepsie Gulch came into prominence in 1879, when Fossett wrote, "Poughkeepsie is a famous mining locality. It contains 250 lodes on which assessment work is done annually, and a large number of lodes are worked steadily—five or ten are paying handsomely."

Taking a short side trip on the Poughkeepsie Gulch Road, there is a large dump visible beside the road after about a mile. A foot trail begins shortly before this point and ascends about one thousand feet to the original site of the Old Lout, one of the principal mines in Poughkeepsie Gulch. The mine was located near the top of the gulch in 1876 and was worked through a shaft, which produced an estimated four hundred thousand dollars, mainly in silver and lead, before closing during the Silver Panic of 1893. It was named after Oliver D. Lotzenhouser. The tale is told that when the shaft was down three hundred feet the miners were told to give up. There was one stick of dynamite left, which the last man decided to use. It opened up a very rich vein and eighty-six thousand dollars was taken out of the mine during the next month. The tunnel at the bottom of the gulch was driven in 1886. The Old Lout has been worked off and on several times since then. The Maid of the Mist is an extension of the same vein.

Farther up the road, to the east of the creek but west of the road, are the remains of a few log cabins that mark the site of the Poughkeepsie Mine. This early mine was located in 1873 (the first year the San Juan Mountains were legally open for mining) by R. J. McNutt. The name of the gulch was changed from Alpha to Poughkeepsie at this time. The mine produced a high grade of silver-copper ore. An early thirty thousand dollars offer to buy the mine was refused because it was located on one of the major veins of

This is the upper workings of the Old Lout. Later a tunnel was driven at the bottom of Poughkeepsie Gulch and the activity shifted to that location. Author's Collection.

the entire San Juans (still easily visible across the creek). It is now a nice spot to camp, but from this point to Lake Como the road is virtually impassable for four-wheeled vehicles. To continue in a vehicle would require tearing up the tundra, traveling over some really big rocks, and navigating some extremely deep potholes.

Almost 150 claims lay within a mile of Lake Como, all of them above 11,200 feet (timberline in the gulch). The ores were rich, but most did not prove to be very substantial and were soon worked out. The many short shafts and discovery cuts in the area of Lake Como belong mostly to the Como Group and the Alaskan Group, whose claims are mainly on the raised bench to the west. The Alaskan Group was said to be the most prominent mine in the district in the 1870s and was sold by Matthew Johnson and P. O. Lunstrom to H.A.W. Tabor (Baby Doe Tabor's husband) and others for $125,000 in 1879. A six-foot vein of gray copper ore averaged eight hundred dollars per ton. Tabor's brother managed the mine. Frank Fossett reported in 1880 (shortly after the purchase by Tabor) that "the mine has been steadily developed by the new owners and new buildings have been erected, including an ore house and quarters for twenty-five men employed." The activity dropped off drastically in 1882 after H.A.W. Tabor's group pulled out.

There were rumors of a lost Spanish gold mine somewhere in the talus slope on the southwest side of Lake Como. An early day prospector, John M. Stuart, reported finding hand-forged copper mining tools in the tunnel in 1878, but soft copper is not a good metal to use for mining tools. No great amount of gold or other valuable minerals were found either, and the whole situation remains an unsolved mystery.

Returning from the Poughkeepsie Gulch side trip to the main Engineer Road and proceeding east for about two rough miles, the Des Ouray Mine is encountered. It

William Henry Jackson caught this family on film in Poughkeepsie Gulch about 1885. The burros carry chairs and other household goods. Courtesy Colorado Historical Society. *(CHSJ769)*

received its name from Des Moines, Iowa, investors. The mine was extensively developed and worked off and on until the 1980s, so there are still cabins, mining equipment, and other buildings visible. The mine is a collection of some twenty claims; the Wewissa, Benach, and the Eurades are the most prominent. In 1925 a small sawmill also operated out of the area.

A mile farther, there is a three-way fork in the road. Coming from the Ouray side, the first choice is a short road that leads across the creek to a pack trail to the southwest. This trail leads through some beautiful backcountry and eventually leads back down to the Old Lout and Maid of the Mist Mines in Poughkeepsie Gulch. Roger Henn in *Lies, Legends and Lore of the San Juans* reports a lost gold mine in the area. James Herring worked at the upper level of the Old Lout and left after work every day to develop a rich discovery, the location of which he kept secret. He died before disclosing the location, and no one has ever found this small but rich mine.

The second choice is a very rough road that goes straight, crossing the creek to east, and heading directly toward the San Juan Chief Mill. There are a few primitive camping spots along this short route, but the road is very rough. The third choice, the main road, curves to the left and ascends through switchbacks to a wooded area where restroom facilities can be found. The "scenic outhouse" affords a good view of the deteriorating ruins of the San Juan Chief Mill. Is the "no diving, shallow water" sign still on the toilet lid? If you stay on the road about a half-mile farther, it is possible to follow another road in the tundra area and then cross back to the San Juan Chief Mill, but the road is rough and there are many little dead-end roads, so be careful and stay off the tundra.

The San Juan Chief Mill was built so that the low-grade ores of the Mineral Point area could be concentrated for shipment from the remote area. Several of its big stamps, which crushed the ore, are still visible. At the site of the mill, two steam boilers also can be seen slowly rusting away. An inevitable question asked by someone on the jeep tours is, "How in the world did anybody get those big boilers up here and why are there two

The San Juan Chief Mill was already suffering from neglect in 1960 when this photo was taken. Author's Collection.

of them?" One of the jeep drivers jokingly explained it this way: "Well, you see, when there was only a pack trail, everything has to be brought up here on pack mules or burros. Loading these animals required that whatever was loaded on one side of the packsaddle had to be counter-balanced on the other side so that the animal wouldn't fall off the trail. They didn't really need two boilers, but they had to bring the extra one along just to balance the load."

The whole area around the mill was called *Mineral Point*, but the actual settlement of that name was about a quarter of a mile across the swampy area to the south of the mill. There is a trail leading out from the San Juan Chief toward the town site of Mineral Point, but because of the tundra and the swamp, the road has been closed to any type of vehicle. It is now necessary to access the town site by hiking or by driving in from the Animas Forks side. The early promotion of Mineral Point was extremely exaggerated, even showing steamships on the Animas River, lush gardens in the area, and a trolley connecting Animas Forks and Mineral Point. It was later estimated that two to three million dollars was "wasted" in such promotions. The area was named for the extensive mineralization and numerous parallel veins that can be seen running on the surface. Ingersoll reported, "Mineral Point, where twenty or thirty rich veins crop up, is covered with claim stakes until it looks like a young vineyard." The summit of Mineral Point contains a heavy, white quartz vein (still visible) that is almost sixty feet in width. However most of the Mineral Point mines were not profitable because of low- to medium-grade ore, high transportation costs, and the extremely remote location that is inaccessible for up to eight months out of the year.

Continuing on the main road east for about a mile from the San Juan Chief, the Engineer Road forks to the left, and zig-zags up the side of 13,218-foot Engineer Mountain. By traveling on to the right-hand (south) road you will end up in Animas Forks. This road was the extension of the original Lake City Toll Road over Engineer Pass, which connected with the Animas Forks and Silverton Toll Road. The road leads first to Denver Lake (there are fish in the lake at times). Across the valley to the east you can see Denver Pass and Horseshoe Lake. The high mountain is Seigal Mountain, elevation 13,274 feet. Denver Pass was used as an early shortcut to Lake City, but it could only be traveled only on foot or by horseback. Now it is a struggle even on foot! The dumps

These 1920s tourists pose in front of what was left of the post office complex at Mineral Point. Author's Collection.

and mine workings on the east side of the road around the lake is the Denver Mine, and the little tin house by the lake was probably used for storing powder. Continuing toward Animas Forks on this side trip, you encounter the Mineral City Road (covered in another chapter) to the right and then in a couple of miles the Cinnamon Road to the left. Then it is just a little over three-tenths of a mile to Animas Forks.

Returning to the Engineer Mountain cutoff, the road progresses through a series of switchbacks up the mountain. About a mile up the mountain a road to the right leaves the main road at a switchback. It is another early day wagon road of about a half a mile that provides spectacular views, but which can no longer be traveled through to Henson Creek. From the overlook at the end of the road you will be looking at (left to right) Wildhorse (13,266 feet), Wetterhorn (14,105 feet), Matterhorn (13, 590 feet), and Uncompahgre (14,309 feet) Peaks and more generally at American Flats and the Big Blue Wilderness area. The road to the top is blocked off but leads leads to "Oh! Point." It was once the main road over the pass.

Back on the Engineer Road, we pass the many dumps of the Polar Star Mine. H. A. Woods first located this claim in March of 1875. He was in Howardsville, near present-day Silverton, and overheard some prospectors talking of going to the spot. Woods had located several mines on Engineer Mountain and knew that the Polar Star had been staked by another man but that the necessary assessment work had not been done. That night, he took off ahead of the others and arrived at the claim at 6:00 A.M. to beat "Sheepskin" Miller out by hours. It was very cold at the time of his arrival, and the presence of the morning star gave rise to the name of the mine. By 1882 the Polar Star was owned by the Crooke brothers and Henry Wood, and the ore was shipped to Lake City. Twenty men worked at the mine, which produced good amounts of silver for many years. The big excitement came in 1882 when a streak of solid silver was found that eventually totaled six hundred pounds.

Near the top of Engineer is a short road leading to a ridge from which there are great views of a good portion of the San Juans. Then a little farther is the official top of the pass (elevation 12,800 feet). The mountain itself is 13,218 feet. Engineer Mountain was first named *Mount Ruffner* for the leader of the 1873 Ute Reconnaissance

Mission, Lt. E. H. Ruffner; but he later asked that the name be changed, and it was renamed to honor the Engineer Corps of the U. S. Army. There are three trails here that are all blocked off. One leads to American Flats, another to the orange-red ridge and its mine, and the third down into the basin that is the headwaters of Bear Creek. Nearby American Flats is an area of thousands of acres of flat and gently rolling country located between twelve thousand and thirteen thousand feet. The area is covered with grass, tundra, and wildflowers in the summer. It is a favorite summer hangout for elk.

Engineer City was located in the basin on the north side of Engineer Pass. The "city" was basically the Frank Hough Mine. The Frank Hough was discovered in 1879 by John Hough (Frank's father) and produced gray copper, copper pyrites, and iron pyrites with silver running fifty to sixty ounces, twenty percent to twenty-eight percent copper, and traces of gold. As many as three hundred men were working at the Frank Hough during the summer in the early 1880s, most of them boarding in tents. Its vein was two to seven feet wide and by the early 1890s it had produced over $250,000. Jack Davidson erected a huge tent which served as a boardinghouse and could hold fifty men. The Lake City *Silver World,* in July 1882, boasted that Engineer City was the only town of its size with no saloon. In the 1890s Lake City touted the Frank Hough as "the largest and richest silver and copper producer in southern Colorado," but by 1900 the mine was shut down and the area basically deserted.

Traveling down through the basin, you meet the trailhead of the Horsethief Pack Trail, which leads through American Flats. If you are hiking in this country it is easy to get lost-be careful! This is a true wilderness, and because the area has no high peaks or trees, it seems as though every year the Mountain Rescue teams go after at least one lost group of hikers.

A few tenths of a mile farther is the remains of a small stone building that was the powder house in the 1880s for the Palmetto Gulch Mining and Milling Company. The shafts and dumps in the area were also part of the mine. About a mile on down the road (after passing through a series of switchbacks to tree line) is the new "Thoreau's Cabin," built in honor of the famous naturalist Henry David Thoreau. Although it is unusual to build such a large and modern private home in such a remote and fragile area—it is a very beautiful structure. The remains of the small Palmetto Gulch Mill are visible across the road to the south, and a small cabin has been preserved across the creek. The Palmetto Gulch Mill processed about twenty-five tons of ore a day with its fifteen stamp mill. Its purpose was to simply concentrate the ore for shipment. The Palmetto Mine produced mainly silver, copper, and iron. In 1882 its ore was running about fifty dollars a ton and by the mid-1890s it had earned over two hundred thousand dollars. It was thought to be the best mine in the district in 1879, but it failed to produce ore that was easy to reduce. The "state of the art" mill only captured about 40% of the valuable minerals in the ore.

Another two miles down the road, at 10,850 feet elevation, is the site of Rose's Cabin (now posted as private property), an early-day stagecoach stop and small settlement. "Rose" was not a woman, but rather Corydon Rose (also called Charles), a man who built a one-room cabin in 1874 and then opened a hotel and bar. Rose sold out to Charles Schaffer in 1878. There was a post office at the cabin between June 1878

This view of the Palmetto Mill about 1905 shows that it was a large operation right on the edge of timberline. Courtesy Colorado Historical Society, A.E. Reynolds Collection. (CHSX4932)

and September 1887. Until 1879 it was the only place for meals, lodging, or liquor between Mineral Point and Lake City. Liquor must have been the main emphasis at the cabin because it was said the bar ran the entire length of the first floor. The cabin also served as a large packing and forwarding operation. The cabin was greatly expanded in 1882. The main hotel was two-and-one-half stories high and consisted of twenty-two tiny rooms. That same year something went wrong at an all-night poker game, and Joe Nevins (evidently a very nasty character) hit Andy MacLauchan in the head with a hatchet. MacLauchan then shot Nevins dead and was later acquitted by a verdict of self-defense. At its peak, the population of Rose's Cabin was approximately one hundred. By 1885 Crofutt described it as "a small mining camp on Henson Creek, 15 miles west from Lake City. The place consists of a post office, smelting works, store, restaurant and about 120 population." In 1920 the barns and buildings at Rose's Cabin were remodeled by the Golconda Mining Company and provided with modern plumbing and the county's first propane tank. The main building was torn down in 1950. Only the partial remains of the hotel's fireplace and foundation and the sixty-mule stable are still visible.

The Rev. George Darley, who established and built the first two churches on the Western Slope of Colorado (the first in Lake City and the second in Ouray), had good reason to appreciate the location of Rose's Cabin on one memorable occasion. It was in the winter of 1890 that the Rev. Darley made a trip on foot from Ouray to Lake City via Engineer Mountain in a heavy snowstorm. He was accustomed to traveling long distances on foot—over, around, and through the San Juan Mountains at all seasons of the year and in all kinds of weather. He left Ouray at 5:00 A.M. in a blinding snow storm, traveling the trail from Ouray up the Uncompahgre River to Mineral Point, then over Engineer Mountain, stopping only once at a cabin where he obtained a warm meal and a chance to warm himself and dry his clothes a bit. At times he forced his way through snow that was waist deep and then even armpit deep as he crossed over the summit of

Engineer Pass. At 9:00 P.M., after sixteen hours on the trail, he gained the sanctuary of the saloon at Rose's Cabin. The miners who were lounging in the warm saloon not only welcomed him, but also congratulated him, for to them it seemed a miracle that any human could have survived that trip over Engineer Mountain in such a storm! Darley later wrote of his adventures in a great book called *Pioneering in the San Juan*. In it he swore that he knew a man who took his burro over Engineer Pass in the dead of winter by putting snowshoes on its feet!

Just below Rose's Cabin was the Bonanza Tunnel and Mill. Mining engineer T. A. Rickard was slightly impressed with the ore in 1892 but was very unimpressed with the new electric drills that the mine was trying out. Rickard mentioned that they would be wonderful to operate if only they would work!

On toward Lake City there is a good restroom facility, and the road is passable for two-wheel drive vehicles. All along this section there can be found camp sites on little dead end roads that lead to the creek, which is usually filled with beaver ponds at this point. There are also several private homes built on mining claims. The first of the small public roads leads not only to some good camping sites but by hiking, you can go across the creek to the Schaefer Mine in the gulch by the same name. Unfortunately, this road is now blocked off from vehicle traffic as a family recently went into the mine and was killed by bad air. Please do not enter any open tunnel because bad air and cave-ins are real possibilities.

About a mile farther the road passes a large mill restored by the Colorado Historical Society. The Empire Chief Mill was constructed during 1927-28 and began operating in 1929. It processed lead, silver, and zinc ore but was hit by an avalanche only two months after it opened and four men were killed. It never fully recovered and by the early 1930s had closed down.

Over the next mile and a half there are some beautiful waterfalls on the other side of the canyon. Then the Rose Lime Kiln is visible in the canyon. George Lee built the kiln in the spring of 1881 to help make cement and mortar for the brick buildings that he envisioned would be built in Capitol City. The lime was also used as a flux for the smelter he owned there. The limestone was mined a little downstream from the smelter at the base of the cliffs above the kiln. The venture was a disaster. Lee only sold six hundred dollars in lime, and the kiln shut down in 1882.

Rose's Cabin in the 1920s was a large structure, but was nearing the end of its use as a hotel/saloon. Author's Collection.

A mile downstream from the kiln is Whitmore Falls. The falls are deep within the canyon and hard to see from the road, but a short trail leads to a good vantage point about half way down to the creek. Several miles down the road you enter a deep aspen forest and then pass livestock pens to the right (downhill).

The remains of the brick smelter (with its beehive dome) are then visible across the creek, and there are private homes to the left. The smelter was built by George Greene (not the same as the Silverton Greenes). Because of its great location, a town had already been laid out and named *Galena City* (it was located in the Galena Mining District) before George Lee arrived on the scene. When Lee bought the two hundred-acre site in 1877, about one hundred people were already living at the location. It was renamed *Capitol City* because the post office already had another Galena City in Colorado. Lee built a two-story brick house in 1879 and called it the Mountain House Hotel. Governor F. W. Pitkin, who owned mining claims all over the San Juans, stayed in the hotel that first year. Lee obviously loved brick, perhaps because of the fire danger in small mining camps, so he also built a brick carriage house and even a brick outhouse patterned after the hotel. He also built the Rose Lime Kiln previously mentioned, two smelters, and a sawmill. When Lee built his brick mansion, some say that he envisioned himself as the governor of Colorado, ensconced in a new Colorado capitol that was to be moved to this site from Denver. Hence the name *Capitol City*. Lee might have misspelled it, but some oldtimers insist the town was originally spelled Capital. The brick mansion contained bay windows, rich paneling, and a small theatre. Frank Fossett said that the Lees were "distinguished for their hospitality." During the big boom of the 1880s Capitol City had about seven hundred residents with a store, several hotels, restaurants, and saloons as well as a schoolhouse and post office from May 18, 1877, until October 30, 1920. Lee's ventures failed during 1882, and he moved away.

Nearby mines include the Capitol City, Yellow Medicine, Great Eastern, San Bruno, and The Morning Star. The Capitol City had a rich vein near the surface but unfortunately it turned to a lower grade ore as it went down. The Capitol City, Lily, and Capitol City Extension were all on the same vein, which was about five feet wide and had a pay streak of about twenty inches of galena, iron, zinc, and copper. Crofutt reported in 1880 that many of the Capitol City mine owners were fighting over their mines and "two smelting works are idle; all waiting for something to turn up." By 1885 the population was down to 120 and by 1895 only fifty. A second boom happened about 1900 when the Ajax and Moro Mines struck large amounts of gold and copper.

The Excelsior Mine is just north of Capitol City at 10,220 feet. It was located in 1878 but didn't ship any ore until 1893. Its ore was fifty to fifty-nine ounces of silver, ten to fifteen percent copper, and five percent lead. The Excelsior, Broker, Silver Cord, and Czar were all probably on the same vein and produced gray and yellow copper ore containing silver and gold and running about $60 to $100 per ton. By 1910 the population of Capitol City had fallen back to about a hundred. Many of Capitol City's buildings were later salvaged for their materials, yet in the 1950s Lee's mansion was complete enough that it could have been restored; but by about 1960, it became necessary to bulldoze the rest of the building to eliminate the danger of people being injured by falling segments of the walls. The old cabin, which still stands, was the post office, restored in 1992 by the Hinsdale County Historical Society.

At the north end of the Capitol City is the cutoff to North Henson Creek. It is a great little side trip that leads to the Matterhorn Trailhead after about two miles and to one of the trailheads to Uncompahgre Peak at the end of the road at about four miles. Along the way are some good campsites. The Capitol City Cemetery is

Lee's Mansion was weathered but still substantially complete when photographed by Muriel Wolle in the late 1930s. Courtesy Denver Public Library, Western History Dept. (X4002)

about three hundred yards up the hill at the start of the road. It is in bad shape but is still interesting. The lone chimney further up the road belongs to the Gallic-Vulcan Mill, which never went into operation. The roads that go uphill all lead to private property (the first to the Czar and Czarrina Mines and the second to the Yellow Medicine Mine and Mill). They are both gated after a short distance. To the south side of the North Fork of Henson Creek were the Julia, Gallic and Vulcan claims. The Vulcan was discovered in 1883 and the Gallic a few years later. Benjamin Guionneau and his relatives did most of the early work. The ore was two-tenths of an ounce of gold, ten ounces of silver, and fifty percent lead with some zinc and iron. The mine had a small concentrator connected by a trestle to the ore house. A little less than three miles farther is Mary Alice Creek; where the road makes a switchback the hiking trail leads to a high basin at the foot of Dolly Varden Mountain. It is best to turn around here because the road ends a short distance farther at a place where it is hard to turn around.

Back on the Engineer Road to Lake City there is a stretch of about ten miles of good fishing and a few camping sites, but there is little of historic interest until you reach the point where a marked road leads a short distance to Snowden Park and the cabin that Pike Snowden built in the late 1870s. He was one of the real pioneers of the San Juans, helping to found the town of Silverton. He lived at this spot until 1915, all the while doing prospecting nearby. The bars on his cabin were installed when he had great success at a poker game in Creede and was afraid that someone might kill him in his sleep to get his money. The meadow around his cabin was a favorite picnic spot for the residents of Lake City. Carolyn and Clarence Wright devote an entire chapter to Snowden in their book *Tiny Hinsdale of the Silvery San Juan.* He loved to tell tales-especially tall tales. The short road near the cabin leads to Snowden's mine. A rough road leads farther up out of Snowden Park, but the same road can also be accessed from a better spot about two tenths of a mile farther along the Engineer Road.

The Nellie Creek Road (Hinsdale County Road 877) is another way to get to Uncompahgre Peak. The Nellie Creek route was traveled by whites as early as the Rhoda party in 1874 when they set up a survey station at the top of the peak. No one seems to remember who "Nellie" was. Uncompahgre Peak at 14,309 feet was at one time thought to be the highest peak in Colorado, but it is now known to be the sixth. It is not a difficult climb, and the top is quite flat. Forest Road 877 travels a little more than four miles through some beautiful country, but there is not much of a historic nature along the route. A beautiful waterfall is about three-quarters of a mile up the creek, and at two miles there is a water crossing. Just after the ford are a couple of switchbacks, and then a road leads for a short distance to the right. It is the access to the east fork of Nellie Creek, where a hike will take you to an old sawmill after a mile and a half and also several small mining prospects. At three and a half miles along the Nellie Creek Road

Capitol City never came close to filling the valley in which it was situated, as this photo near its peak makes clear. Courtesy Colorado Historical Society. *(F40176)*

is good camping and a trail to the headwaters of El Paso Creek. At about four miles are the Uncompahgre Trailhead and good restrooms. The hiking trail is the shortest way to the popular peak. Allow three or four hours for the hike. This also marks the beginning of the Big Blue Wilderness.

Back on the main road, after about three-tenths of a mile, there are good restroom facilities. A few small mines were along this route, including the Big and Little Casino, Scotia, Pearl, Fairview, Vermont, and Alabama. The Ocean Wave (eight miles from Lake City on Red Rover Mountain) mined a five-foot vein of galena and gray copper. In 1877 it was described as the best developed mine in the district. A lot of the road work into Lake City was done to get to the Ocean Wave, because it had a large mill and concentrating works at Lake City. In 1879 Fossett reported that the Dolly Varden Mine along Henson Creek (on Copper Mountain) "carries a very rich vein often four to ten inches." It was yielding $225 to $1,150 per ton from its earliest days and then was sold to the Crooke brothers.

Five and a half miles toward Lake City from Capitol City you reach the Ute-Ulay Mine and the site of the settlement of Henson. Even though the road runs right through the middle of the mine and town, please respect the "no trespassing" signs. The creek and town were named for Harry Henson, a prospector who was in the area in 1871— two years before it was legal to be in Ute territory. Although Harry Henson, Joseph Mullen, Albert Meade, and Charles Goodwin discovered the Ute and Ulay in 1871, it could not be officially located until 1874. The Ute and Ulay are two separate claims that have been worked together almost from their discovery. (The Ulay is closer to the road than the Ute.) Ute-Ulay ore ran twenty to ninety ounces of silver and forty percent lead. There were also large quantities of gray copper. A post office was established at Henson on May 7, 1883, which continued until November 30, 1913. In 1910 a hundred residents were living at Henson.

The Ute-Ulay Mine and the town of Henson about 1885. The structure in the front is the flume to bring water for the mill. Courtesy Denver Public Library, Western History Dept. Frank E. Dean Photographer. (X61914)

The original discovers of the mine sold out for $125,000 in 1876 to the Crooke brothers, who built a reduction works at the site. For a short time afterwards the settlement was called "Crooke's." Originally, their plant consisted of a fifteen-stamp mill and concentrator. In 1878 a lead smelter and chlorination mill were added. Their concentration mill was extremely important to the economics of the mine, because the ore had to be transported to the D&RG Railroad at Gunnison by wagon at a cost of twenty-five to thirty dollars per ton. Part of the ground tramway or "incline" that carried the ore to the mill is still visible. Three tram cars at the top were tied together, loaded with ore, and sent down the rails. Their weight pulled three unloaded cars back up to the mine. The smelter also processed ore from the Polar Star Mine near the top of Engineer. The Crooke Brothers sold the Ute-Ulay Mine in 1880 for $1.2 million. Then the mine was sold again in 1882 to English investors.

In 1883 the Ute-Ulay was reported to be "the most extensively developed mine in the San Juans." Its ore ran sixty to seventy percent lead, fifteen to fifty ounces of silver and about one-third ounce of gold. But the Crooke Mill closed that same year because of the high costs of fuel, flux, labor, transportation, and problems treating the ore. By 1886 even the Crookes' home company in New York had been sold to pay debts—a sad ending for a company that was described by the Lake City paper as the "chief engineer of the city's prosperity." During the 1880s and 1890s, between two hundred and three hundred men were always employed in the area. A terrible accident occurred when a miner in the Ute-Ulay accidentally drilled into the Hidden Treasure (they ran parallel to each other), which caused a gas explosion and killed twenty Ute-Ulay miners and sixteen miners from the Hidden Treasure. Total production of the Ute-Ulay Mine eventually exceeded twelve million dollars, making it one of the great mines of the San Juans.

Clark's Studio took this photo of the abandoned smelter at Granite Falls in 1892. Courtesy Denver Public Library, Western History Dept. (X61418)

The production records of the Ute-Ulay and its promised high-tonnage shipments were the most persuasive arguments to get the D&RG Railroad to build a branch line from the main-line at Sapinero (now under the waters of Blue Mesa Reservoir) to Lake City. Unfortunately, a slump hit the local mining industry shortly after the railroad arrived in 1889, followed by the Silver Panic of 1893. Shipments of Lake City ore declined year by year after the railroad was built. If the lack of good ore wasn't bad enough, there was a major labor struggle at the Hidden Treasure and Ute-Ulay Mines on March 14, 1899. Eighty Italian miners, who belonged to the United Federation of Miners (UFM), struck over the requirement that all single men had to live in the boardinghouse. The Italians met the day shift that day and wouldn't let them into the mine. Some nonunion men were beaten. It was then discovered that fifty rifles and one thousand rounds of ammunition were missing from the armory in Lake City. The Italians also bought almost every gun and box of ammunition in town. The sheriff arrested twelve Italians, charging the secretary of the union with breaking into the armory. By March 15 the sheriff was convinced that the Italians would become violent and telegraphed the governor, who sent 326 militiamen to Lake City. When the Italian consul from Denver talked the miners into laying down their arms, the Italians were all fired, and the troops left on March 20. The Ute-Ulay never again hired Italian miners and work resumed at the mine, but the missing guns and ammunition were never recovered.

In the early 1920s the great mine closed down, but in 1925 Michael Burke, a wealthy mining promoter, bought the mine, brought in modern mining equipment, built a new 100-ton mill, constructed a sixty-foot-high dam on Henson Creek downstream from the mine, and installed a hydroelectric generating plant. It was Burke's influence that prevented the abandonment of the Lake Fork branch of the D&RG for several years; Burke claimed that all his expenditures on the Ute-Ulay would be for naught without the means of shipping his product by rail. However, the 1929 Depression hit, and metal prices dropped so low that Burke was never able to make the ore shipments that he had promised. In 1932 the D&RG Railroad finally obtained the consent of the Public

Utilities Commission to abandon the line. Burke then bought the branch line and tried running a railroad himself, just for the sake of his Ute-Ulay Mine, but in 1937 he was obliged to give up railroading as well as operating the mine.

The Hidden Treasure Mine is three miles above Lake City or about a mile down from the Ute-Ulay. It had a mill and a small town called Treasureville. The mine was discovered in 1874, but not much work was done until 1890. It did very well from 1897 to 1930 (producing about $1.2 million). The Ute-Ulay and Hidden Treasure were combined in 1940, and in 1952 a new dam and mill were constructed.

A mile and half farther toward Lake City are the remains of the Pelican Mill. Only the foundations remain now. The mine was located high above the mill. Across the creek was the Fanny Fern Mine. A little more than two miles downstream at the junction of Henson Creek and the Lake Fork of the Gunnison River lies the town of Lake City. One could, of course, return by the same road, but it is also possible to travel back to Silverton or Ouray over Cinnamon Pass.

The Cinnamon Pass Road was one of the earliest in the San Juans. A trail over the pass was used for centuries by the Ute Indians, later by Charles Baker who came into the San Juans over this trail in 1860, and also by Franklin Rhoda of the San Juan Division of the Hayden Survey in 1874. That same year Enos Hotchkiss built a toll road for Otto Mears from Saguache (where Mears lived) over Cinnamon Pass to the new settlement of Silverton. Good mineral discoveries were made during its construction, and the town of Lake City was therefore developed. However the pass was rough and was never really passable to wagons, so it fell into disuse in the early twentieth century and was closed to traffic for many years before it was reopened as a jeep trail in the 1950s.

As you start out of Lake City, Crooke Falls is about one mile up the road.. The seventy-five-foot drop of Granite Falls powered the fifteen-stamp Crooke's Reduction Works, which was established in August of 1876. The Crooke brothers also purchased the Ute-Ulay Mine that year. Other crushers and a smelting works were added in 1878. Coal and coke were brought in all the way from Crested Butte by wagon. John Crooke was a major mining investor who got much of his money from his invention of tinfoil. That first year the Crooke Mill processed $85,498 in silver, $23,698 in lead, and $2,925 in gold. Crofutt reported that near the mill "quite a village has grown up, with stores, hotels, saloons, etc., all of the residents being employees of or dependents upon 'the works.'" The Crookes also owned claims just north of the Golden Fleece. About two miles out of town is the Gold Coin. On the other side of the valley at this point were the Belle of the West and Belle of the East Mines-both following the same vein.

At the fork that is a little over two and a quarter miles out of town, the main road (Highway 149) leads up Slumgullion Pass (elevation 11,361 feet) on the north side of the lake and then continues to Creede. Both of the early toll roads went out of Lake City over Slumgullion Pass but after that point one turned south over Spring Creek Pass and on to Del Norte while the other went northeast over Los Pinos Pass to the Cochetopa Pass road and down to Saguache. Crofutt called Slumgullion a "villainous mountain." The toll road had many corduroy sections (logs laid into the road) that Crofutt said gave the effect of "walking backwards up the stairs, and then sliding down with the feet slightly elevated.... The dirt when wet seems to have no bottom and will stick as close to anything as some hackmen to a 'tenderfoot.'"

Alferd Packer, the man-eater, was sentenced to death, escaped, and because of a legal technicality sentenced to life in prison. Courtesy Colorado Historical Society. (F11958)

The Slumgullion Pass road is a side trip worth taking. It follows the approximate route of the toll road. About half a mile up the Slumgullion Pass road one can find the Alferd Packer monument-dedicated to the memory of the slain men who were a part of one of the few cannibal incidents in U.S. history. There is still a debate over whether Packer's first name was spelled "Alferd" or "Alfred," but he actually preferred to be called "Al." The five men with him were Israel Swan, George Noon, Shannon Wilson Bell, James Humphrey, and Frank Miller. The group had two rifles, a skinning knife, and a hatchet between them. Packer and his five companions left Provo, Utah, on November 8, 1873, and tried to make it across the San Juan Mountains in the dead of winter, even though they had been warned by Chief Ouray not to make the trip. The group became lost, and Packer came into Saguache alone sixty-six days later. He didn't seem to be in bad physical shape and had a lot of cash. Packer gave three versions of what had happened—he originally said he had been separated from his party and had no idea where they were. When a search party was organized and found the rest of the group's partially eaten bodies, he told the last two versions—Bell killed the men and after slaying the killer Packer ate them, or the other men died of hunger and Packer ate them. There are present-day attempts to use forensic science to determine the truth of who killed whom, but we may never know the real truth, because Packer maintained his innocence to the end. The exhumation did show that almost every piece of flesh had been hacked off their bodies and that the men met a very violent end. One thing is admitted—Packer ate his companions!

Packer was arrested but escaped and remained at large for nine years before being captured in Wyoming. He was convicted of murdering the other men by a Hinsdale jury and sentenced to death. At the time it was reported that Judge M. B. Gerry (a Democrat) sentenced Packer and included a statement that he should be hung because "there were only seven democrats in Hinsdale County and you ate five of them." The tale was actually made up by saloonkeeper Larry Tolan, but it makes a good story. The Colorado Supreme Court later reversed the conviction and ordered a retrial. This time Packer was convicted of five counts of manslaughter and sentenced to forty years in prison. He ended up being paroled after serving fifteen years and died of natural causes in 1907.

Back to the west in Deadman's Gulch is the Golden Wonder Mine. A young boy who picked up some "pretty" rocks (which his employer recognized as gold) supposedly discovered it. Fortunately, the boy was able to take his employer back to where he found the rocks. At one time the mine was considered to be the "richest gold property in Colorado." It may still be because it is one of the few gold mines still being worked. Some of the ore that the mine shipped ran over forty thousand dollars per ton. Nearer the road were the Belle of the West and Belle of the East. They produced over one hundred thousand dollars in both gold and silver.

About a half mile up the Slumgullion Road, the Sawmill Road leads to the right. It is a favorite of locals for snowmobiling. Sawmill Road is named for a major sawmill that operated in the park for many years. Then two more miles further is the scenic overlook. It affords a great view of Lake San Cristobal, which is the second largest natural lake in Colorado. The lake is three miles long and a mile wide at its widest point. Lt. E. H. Ruffner named it for a fictional spot in one of Tennyson's poems. Ruffner was reconnoitering the area in preparation of the removal of the Ute Indians and the opening of the San Juan Mountains for prospecting in 1873. Rhoda, in 1874, called the lake "by far the finest of the many little lakes we saw during the summer." His party arrived at the lake the day after *Harper's Weekly* illustrator, John Randolph, discovered the skeletons of Packer's victims.

Today, the lake provides good fishing and boating, which helps the community of Lake City remain a popular resort, even though it is a considerable distance from larger communities. The Golden Fleece Mine (also called the Hotchkiss Lode) can been very clearly seen across the lake. It was one of the great mines of the San Juans and is discussed a little later along the Cinnamon Pass Road. Just two tenths of a mile farther on the Slumgullion road is a Forest Service trail, and then another mile farther is the Slumgullion Earthflow Overlook and good restroom facilities. The road continues on to Creede, but we will turn around at this point and return to the fork that leads to Lake San Cristobal.

One hundred sixty foot Argenta Falls can be seen just about a mile past the junction of the Slumgullion and Cinnamon Pass roads. You will know when you are passing through the bottom of the Slumgullion Slide when you see the yellowish, sandy soil, and at a little over a mile is the lake itself. Lake San Cristobal was formed six hundred to seven hundred years ago when the huge Slumgullion Mud Slide descended into the valley and dammed the Lake Fork of the Gunnison River. The origin of the slide's name is in dispute. In one version it is named for an early miner's stew that was yellowish in color and didn't smell too great- just like the pass. But another is that it reminded New Englanders of the discarded entrails of a whale being cut up for blubber. It is also, of course, possible that the stew got its name from the whale's entrails. The flow starts at about 11,400 feet on the western edge of what was later named Cannibal Plateau. The original slide has stopped, but there is an overlying flow that is still moving slowly toward the lake.

Driving south along the shore of Lake San Cristobal, one encounters the lower levels of the Golden Fleece Mine, which was discovered in 1874 by Enos Hotchkiss. Hotchkiss found no extremely valuable ore and abandoned his claim, which he had called the "Hotchkiss." George Wilson and Chris Johnson later relocated it under the name

Golden Fleece. Although some pockets were very rich, the early ore from 1875 averaged $150 to $180 per ton. In the 1880s there were pockets of richer ore discovered, but it was a hit-and-miss situation. By 1885 Crofutt reported that on the average the Golden Fleece produced one thousand dollars a ton ore.

Only small pockets of such rich ore were found until 1889, when the value of the ore picked up significantly. Then in 1891 there was a large amount of valuable tellurium gold ore discovered. Charles Davis sold the mine in 1891 for seventy-five thousand dollars. Some people said the new owners had been swindled. However by 1895 ore shipments averaged twenty four thousand dollars per month. In 1896 the mine was working a hundred men and shipped nine railroad cars of ore that each brought between $33,000 and $49,500. At today's prices that would be about a million dollars a car! In just a few months the mine shipped $1.6 million in ore (worth close to $25 million today). The richest ore of the Golden Fleece was reported to run twenty-five percent gold and forty-five percent silver and was shipped directly to the smelters. The lower grade ore was milled near the lake. The mine continued to produce good gold values for many years to come and became one of the great mines of the San Juans. Altogether the Golden Fleece shipped fifteen million dollars in ore worth perhaps three hundred million dollars today. There was a settlement nearby called *Lakeshore.* It was small but had a school, stage service, and a post office from 1896 to 1904.

A mile farther, along the edge of the lake, is the public boat launch, and then at about another mile and on the right, is the Red Mountain Campground, which has restroom and camping facilities. Lake San Cristobal itself is very cold because the elevation is 8,995 feet, but the fishing is good! About three-quarters of a mile further is the end of the swampy area at the upper end of the lake, which is a good place to look for moose (be careful they have a nasty temperament). The bridge over the river takes you to the road that passes around the other side of the lake. There is a Forest Service Campground (Wupperman) along this road. This is also the spot where the paved road ends.

Following the Lake Fork of the Gunnison River for about four miles past the lake (the stream provides excellent fishing) brings us to Camp Red Cloud and a quarter of a mile farther to the Williams Creek Trailhead and then the Williams Creek Campground (a fee area). That's the Continental Divide to the left (south). The Williams Creek Trail can be followed for many miles around Grassy Mountain (elevation 12,821 feet), and eventually it ties back in to Alpine Gulch Trail. Shortly past the Williams Creek Trailhead and across the river is the Camp Trail, which leads five miles to the Continental Divide, to the Colorado Trail, and then over to the Rio Grande River drainage.

A little more than two miles beyond the Camp Trail on the Cinnamon Road is a four-wheel-drive road that leads to the left off the main road and up Wager Gulch to the ghost town of Carson, which sits atop the Continental Divide. At one time the area at the beginning of the road was called *Child's Park* and had a post office, but it was just a stop for mail and supplies going to Carson City and not a real town. The Carson road (County Road 36) can be a little hard to find, and it is also somewhat difficult to travel in wet weather because of mud and clay. The road is about a mile past Castle Lakes and three-quarters of a mile past the Old Carson Bed and Breakfast. You

This William Henry Jackson photo of 1874 shows Lake San Cristobal before the natural dam was raised, making the lake much bigger. Courtesy Denver Public Library, Western History Dept. (Z3452)

will know you are on the right road when it crosses Wager Creek (after a little less than a mile) and winds upward through heavy aspen and then fir and spruce forests. About three miles from the start the road levels out into a large, broad valley. About a mile and a half farther the road forks. The left road leads to Carson City, which is visible from the fork. It was also called *Bachelor Cabins*, because it served the Bachelor Mine, but it is generally referred to today as Carson or New Carson.

Carson City is one of the best-preserved (and partially restored) ghost towns in all of Colorado because it is little known and hard to find. The livery stables still have stalls, there is an outhouse, and there are many cabins. It was established in 1882 and named for J.E. Carson (not "Kit," but a nephew), who discovered ore in the area in 1880 and staked the Bonanza King and several other claims in 1881. The original discoveries sat right on top of the Continental Divide, but within a few years mines were spreading down each side. The lower camp on the Lake City side was based on gold discoveries while the very top and the Creede side were based on silver. The main producers were the St. Jacobs Mine (which is credited with production of over one million dollars in ore) and the Bonanza King. However, there were more than a dozen prosperous mines and 150 claims within a mile or two of the town. Carson had a post office from September 1889 to October 1903. Most of the mines surrounding the town produced lead, copper, gold, and zinc as well as silver and gold. The first road to Carson City was built up Lost Trail on the Creede side of the divide in 1887. The 1890s were a good time for Carson, until things slowed down after the Silver Panic of 1893. After 1896 there was mainly gold mining on the Lake City side of the pass. About four hundred people lived in Carson between 1900 and 1902, but soon thereafter the mines went downhill sharply and by 1910 the population had fallen to twenty.

The right fork of the main road at Carson City will take you to the Continental Divide after a mile and then over into Old Carson. After 1920 most of the ore that came out of Carson was shipped over the pass to a smelter in Del Norte. The Maid of Carson (part of the St. Jacob's Mine) had a vein seven to eight feet wide with up to sixteen hundred ounces of silver and eighteen ounces of gold. Other mines included the Legal Tender, Kit Carson, Iron Mask, and the St. John's Mine. The St. Jacob's Mine straddles the Continental Divide.

The crew of the Golden Fleece didn't take the time to dress up before this photo was taken about 1880. Courtesy Denver Public Library, Western History Dept. (X60892)

Shortly after passing over the divide, the road has three forks. The right fork ends after a quarter of a mile but has some spectacular views of the Rio Grande drainage. The Colorado Trail, which traverses Colorado from one end to the other along the Continental Divide, can also be intersected off this road. The middle fork dead-ends into the ruins of Old Carson. The left fork takes you farther up along the eastern side of the divide and then down the Heart Lake Road and all the way to Creede if you want to go there.

Returning down to the main Cinnamon Pass Road, there is a restroom a little more than a mile farther and also some primitive campsites. There is also fishing access to the Lake Fork of the Gunnison at Bent Creek. The Mill Creek Campground is a little less than a mile further upstream, and a mile and a half beyond that point the lower fork in the road leads for a half mile to the site of the former town of Sherman. The town is now a little hard to find because of flooding and the new growth of the forest.

Sherman was established in 1875 at the junction of Cottonwood Creek and the Lake Fork of the Gunnison. It was platted in 1877 by A. D. Freeman and others, but it was never officially incorporated into a town. Sherman contained several stores, a sawmill, saloons, the Sherman House Hotel (which opened in 1881), and the large Black Wonder Mill, which dominated the town. Crofutt spoke highly of the Sherman house-a hotel, which also contained "a store full of merchandise (which) tempts the 100 citizens to spend their money at home.... This is strictly a mining camp of both placer and lode mines." A post office was operated at Sherman from 1877 to 1898. The summer population probably reached a maximum of about three hundred but only forty or fifty hardy souls stayed for the winter. Crofutt gave it a population of 100 in 1880.

Some of the best mines in the area were the George Washington, New Hope, Mountain View, Minnie Lee, Clinton, Smile of Fortune, and the Monster. Although the Black Wonder ore was originally announced to assay at $167 of gold and $17,174 in silver per ton, the ore either declined rapidly in value or someone misread his figures. The Black Wonder was the main mine in the Sherman area. It produced gold from copper pyrites and silver in ruby and brittle form. A tram was built from the mine to the new mill at Sherman in 1895. The mill was moved in from Whitecross and was originally designed specifically to recover gold, but unfortunately the Sherman ore contained mostly silver. The mill was redone in 192, and the town again prospered for a short time. What little remains of the Black Wonder Mill is located near the river. The flooding of Cottonwood Creek caused the mill's dam to break and the Lake Fork of the Gunnison has obliterated most of the town site, although the remains of several cabins or cabin foundations are still visible. The last mining near Sherman was in 1925.

Several beautiful new homes and rental units have been built around the Sherman town site, and there are a few primitive campsites. You can continue past Sherman and follow Cottonwood Creek for several miles on a rough jeep trail. There was a small settlement along the creek called Garden City or Gardner City. The fishing is often good, and the trailheads to Cataract Gulch and Cuba Gulch are just a short distance up the creek. Both are rough, high mountain hikes, so be prepared with good maps and plenty of provisions.

Back on the main Cinnamon Pass road, after journeying about a mile you will come to the Sherman Town site overlook—a good place to get a view of the whole valley and learn some more information about Sherman and the local mines. The Black Wonder Mine is on the right of the road (look for its dump in the trees). The road then continues along a shelf blasted from the solid rock for a little more than two miles before reaching Burrows Park. The "Shelf Road" contains a tight turn where the mill's water flume can still be seen hanging from the cliffs (it's not a bridge). Burrows Park offers excellent views of Red Cloud and Sunshine Peaks (14,034 and 14,001 feet, respectively). Franklin Rhoda and Ada Wilson (the topographer of the party) ran into a massive electrical storm while on the top of Redcloud Peak. Their hair stood up on end, and the rocks gave off a buzzing sound like bacon frying. They leapt off the peak just seconds before a lightning strike hit where they had been standing. In the early days Redcloud was also referred to as *Red Mountain* and Sunshine was known as *Sherman Mountain*. After crossing Silver Creek you reach the Silver Creek Trailhead, which will take you to the top of both fourteeners. There are also good restroom facilities here. Across the West Fork of the Gunnison to the south is the Grizzly Gulch Trailhead, which leads up 14,048 foot Handies Peak (named for an early surveyor), but a shorter and easier route is from American Basin, a little further along the main road.

Within a two-mile distance, at the west end of Burrows Park, were the towns of Whitecross, Sterling, Argentum, Tellurium, and a settlement called Burrows Park. Considerable confusion therefore exists as to the location of each. Burrows Park is named after Albert Burrows, who prospected in the area in the early 1870s. The spot near the trailheads is generally conceded to be one of the locations of the town of

The Black Wonder Mill is behind the cabins at Sherman about 1895. Courtesy Colorado Historical Society. Harry H. Buckwalter Photo. (CHS-B305)

Argentum (also called Burrows Park). It was located about November 1876 in the flat area north of Copper Creek. It had a couple of hotels, two stores, a blacksmith shop, and a post office as well as a dozen cabins. The smaller of the two cabins that still remain on the site was the post office and was built in the 1870s. The larger cabin was the stage stop and hotel. The Hinsdale County Historical Society restored these cabins in 1989. In 1881 Crofutt placed the population of Burrows Park at fifty to one hundred. In 1885 Crofutt noted the population to be "from 10 to 100." He felt the best mines in the area to be the Undine, Napoleon, and the Ouida. He also noted that some people called Burrows Park *Argentum* or *Tellurium* and that no regular stage or conveyance ran through the area. Burrows Park had postal service from September 1876 to September 1882, although the mining camp was generally called Argentum. Then to confuse matters further the post office was moved to Whitecross, a mile or so down the road.

The post office remained at Whitecross until 1887. The town was named after Whitecross Mountain, which is located across the canyon and tops out at 13,542 feet. It in turn was named for the small white cross of quartz located in the right side of the flat knob near the top of the mountain. The town was first located in 1880. The Hotel de Clauson was a popular part of the camp, and there were several other businesses. Only a few of the men actually lived in the town, most living or working nearby at mines further up the road. The population of Whitecross fell drastically in 1895 when its mill was moved to Sherman, but rose again when the Tabasco Mill was built in 1901. It was at this time that it reached its peak population of about three hundred with a store, hotel, saloon, and large boardinghouse.

Just a short distance above Whitecross was the "town" of Tellurium. It was founded in 1875 near the junction of Adams Creek and the Lake Fork, and by 1878 people were

The little town of Whitecross about 1915. Whitecross Mountain is in the background.
Courtesy Denver Public Library, Western History Dept. George L. Beam Photographer.
(GB7369)

spending the winter at the camp. In 1880, when Crofutt visited the town, he referred to as just "a small camp of a dozen persons." It was named in the hopes of finding the gold ore found elsewhere in the San Juans, but unfortunately not found here. Tellurium and Whitecross were close enough together that the mail could go to either place. In 1900 Whitecross had eleven cabins to Tellurium's two. The cabins were scattered over a half-mile on both sides of the road, but there was one place where there were about five cabins close enough together to be considered "a town." Tellurium gained the post office from 1887 to 1912, mainly because it was closer to the mines that were operating further up the canyon. Somewhere above Whitecross was the short-lived settlement of Sterling, which was probably little more than a tent city.

After leaving Burrows Park at the upper end of Tellurium, there is a large mine visible across the creek to the east. The remains of several cabins, boilers, dumps, and some mining equipment are visible. Then about a half mile further is the road leading south into American Basin, which is a wonderful place to view the high country's summer wildflowers. There are several small mines in the basin, and at the end of the road is the other end of the trail leading to Sloan Lake and the top of Handies Peak.

The main road then starts a steep ascension toward Cinnamon Pass. After a little over a half-mile the road passes the ruins of the Tabasco Mill. The mill was connected by an aerial tramway with the Tabasco Mine, which is about two miles above and just past the crest of Cinnamon Pass. The Tabasco Sauce Company owned the mill and mine. The Tabasco and nearby Premier Mines started producing well in 1898. The mill was built in 1901 and received its power from an electric plant in Sherman. It was one of the first to use alternating current for power. It had a one hundred-ton capacity but operated for only four years. The Bon Homme and the Champion Mines were also worked in the area.

A little less than two miles further, from the top of Cinnamon Pass at an elevation of 12,640 feet, a fine view can be had back to the northeast of 14,050-foot Handies Peak, with Redcloud Peak (14,017 feet) behind and Whitecross Mountain between the two. The remains of the various tunnels of the Tabasco Mine are scattered around the top of the pass, as well as the Isolde and Bonhomme Mines. The road then descends (on San Juan County Road 5) along Cinnamon Creek and into the valley of the Animas River. Please stay on the road because this is a fragile tundra area. The basin used to be a favorite for sheep herds, and the animals could often be seen strewn across the valley like large white rocks. A sheepherder's wagon and several of his faithful dogs were always nearby. There was concern that the sheep were damaging the tundra (and they certainly kept the wildflowers mowed down) so they are no longer allowed in the area. The Baker Party crossed this divide in 1860 and supposedly gave the pass its name for the reddish grasses that carpet the slopes headed down toward the west. When the San Juan Division of the Hayden Survey passed over Cinnamon in the summer of 1874, traveling toward what they called "the far-famed Baker's Park," Rhoda mentioned, "How the people of Saguache ever expect to bring a wagon road up this, I cannot see."

In two and a half miles you reach the site of Animas Fork (elevation 11,200). A dozen houses, the unique wooden jail structure, and the foundations of the Gold Prince Mill still remain to attest to the busy community that was the terminus of the Silverton Northern Railroad. There are good restroom facilities near the ghost town.

MAP OF SILVERTON TO ANIMAS FORKS ROAD

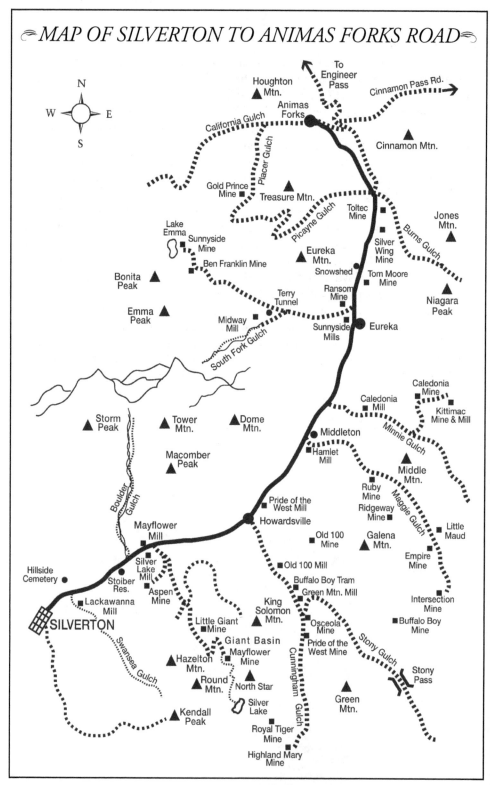

N
W E
S

To Engineer Pass
Cinnamon Pass Rd.
Houghton Mtn.
Animas Forks
California Gulch
Cinnamon Mtn.
Placer Gulch
Gold Prince Mine
Treasure Mtn.
Toltec Mine
Jones Mtn.
Picayune Gulch
Burns Gulch
Lake Emma
Sunnyside Mine
Silver Wing Mine
Ben Franklin Mine
Eureka Mtn.
Snowshed
Tom Moore Mine
Bonita Peak
Ransom Mine
Niagara Peak
Emma Peak
Terry Tunnel
Midway Mill
Sunnyside Mills
Eureka
South Fork Gulch
Caledonia Mine
Caledonia Mill
Kittimac Mine & Mill
Storm Peak
Tower Mtn.
Dome Mtn.
Minnie Gulch
Macomber Peak
Middleton
Hamlet Mill
Middle Mtn.
Ruby Mine
Boulder Gulch
Pride of the West Mill
Howardsville
Ridgeway Mine
Maggie Gulch
Little Maud
Mayflower Mill
Old 100 Mine
Galena Mtn.
Empire Mine
Silver Lake Mill
Old 100 Mill
Hillside Cemetery
Stoiber Res.
Aspen Mine
Buffalo Boy Tram
Green Mtn. Mill
Intersection Mine
Lackawanna Mill
King Solomon Mtn.
Osceola Mine
Buffalo Boy Mine
SILVERTON
Little Giant Mine
Pride of the West Mine
Swansea Gulch
Giant Basin
Cunningham Gulch
Stony Gulch
Stony Pass
Hazelton Mtn.
Mayflower Mine
North Star
Round Mtn.
Green Mtn.
Silver Lake
Kendall Peak
Royal Tiger Mine
Highland Mary Mine

⌐ *Chapter Eight* ⌐

SILVERTON TO ANIMAS FORKS

When traveling towards Animas Forks from Silverton on State Highway 110, you are basically following the path of Otto Mears's Saguache to Silverton toll road and later the route of the Silverton Northern Railroad. As mentioned earlier, the toll road was one of the very earliest routes into the San Juans. The railroad, however, was one of the last little narrow gauge railroads to be built. Otto Mears started the toll road in 1873 and had it pretty well completed by 1875, although the higher part over Cinnamon Pass was only a trail. The actual railroad construction was not begun until 1889, and then only as an extension of the Silverton Railroad to the Silver Lake Mill—a little over two miles from Silverton. The Stoiber family paid for much of the cost in exchange for discounted freight costs. The Silverton Northern Railroad was organized in September 1895 when Mears' new little narrow gauge railroad was extended to Howardsville, then to Eureka in June 1896, and on to Animas Forks in 1904. The Silverton Northern operated until January 1939 when Eureka's Sunnyside mill closed (the Eureka-Animas Forks portion of the railroad had been torn up in 1923).

Before actually leaving Silverton, you might venture onto Kendall Mountain to the southeast of town. There is a four-wheel-drive jeep road that leads seven miles up the mountain, which is also used for one of the premier mountain foot races in Colorado. This route also makes a good hike. The race was inspired by a saloon bet in the old days over whether a miner could run to the top of the mountain and return in less than two hours. Several men did it.

After you leave the city limits of Silverton to the north and start up the paved road, the Lackawanna Mill is located almost immediately across the Animas River to the southeast. The mill was constructed in the 1920s and was connected by a tram to its mine in Swansea Gulch on Kendall Mountain. The Lackawanna Mine ran fifty ounces of silver and twenty percent lead in 1883. The Lackawana Mill also treated ore from several other mines in the Silverton area before it closed in the 1960s. A hiking trail goes from the Kendall Mountain Community Center to the mill.

To the left on Highway 110 (after passing the Cement Creek Road that leads six miles to Gladstone) is the beautiful Hillside Cemetery, which is located on a small hill overlooking Silverton. The first burial in the cemetery was in 1875—the young daughter of Mason Farrow. Approximately twenty-six hundred people are now buried here—well over a hundred died from avalanches, and many more were killed in mining accidents.

Tailings now cover most of the area to the north of the road up to the Mayflower Mill. Tailings are the residue left when ore from the mine is processed at a mill. Tailings are barren, often like quicksand, and reclaiming them with any type of vegetation is a daunting task. This site is an excellent example of the reclamation that is being done by many of the larger mining companies. The Mayflower produced about ten million tons of tailings over its half century of operation.

In 1883 Le Meyne City was started here at the mouth of Boulder Gulch. "Boulder Gulch Camp" was located at the spot before that time. There were six or eight

The large and elegant Stoiber mansion (Waldeim) sat right by the mine's power plant. Courtesy Denver Public Library, Western History Dept. (X-62272).

buildings, and a plat was even filed, but Le Meyne City never took hold, and the site is now buried under the Mayflower tailings pond. A hiking trail from the cemetery will take you up to Boulder Gulch, where there is normally a spectacular wildflower display in the summer.

A little over a half-mile further, to the south of the road, is the three-story brick power station built by Animas Power and Light in 1906. Later, the Sunnyside Gold Corporation and Standard Metals used it as a warehouse. It is now in the process of being restored.

A little less than a half-mile beyond the substation is a nice view area and interpretive sign overlooking the site (across the river) of the former Edward Stoiber home (officially called Waldheim). It was an enormous three-story structure built in 1897 and containing thirty rooms. The mansion was close to the Silverton Northern Railroad tracks (now the roadbed you see running next to the river) and within easy walking distance of the Silver Lake Mill. A large, pillared verandah wrapped around the front of the house, and inside the mansion the ceilings were twelve feet tall. The house included a ballroom, theatre, conservancy, and dozens of bedrooms. Right next to Waldheim was the coal-fired power house for the mine, mill, and, of course, the house. Ed Stoiber and his brother Gustav were experienced mining men who came to Silverton not as prospectors but as capitalists. They set out to buy mines, open a smelter, and start production as quickly as possible.

Stoiber's wife, Lena (appropriately nicknamed "Captain Jack"), ran the house as a stern captain might run his ship, which perhaps explains why Mr. Stoiber spent much of his time at his Silver Lake Mine. Lena caused enough friction between Ed and his brother that it became necessary to split their mining properties. Ed took over the Silver Lake Mine and other claims, while Gustav retained the Iowa-Tiger and several equally rich claims nearby. It was after the breakup that Waldheim was built.

Lena ruled Waldheim with an iron fist. It was said she had mirrors installed so she could watch several of her maids at the same time. It was reported that she favored paying the miners at the Silver Lake only twice a year-on July Fourth and at Christmas. She reportedly took her wagon downtown, get her men out of the bars, and take them back to work. Lena drove the ceremonial last spike on the Silverton Northern at Eureka in 1906.

In 1902 the Stoibers sold the Silver Lake Mine and their other holdings (175 mine, mill, and placer claims) to the American Smelting and Refining Company for $2.3 million and moved to Denver. Ed later died in Paris. Lena and her third husband were on the maiden voyage of the Titanic when it went down. He died in the tragedy, but Lena survived along with the more famous Molly Brown. For many years the mansion served as the manager's quarters for the Silver Lake Mine, but for many more years it was the living quarters for the watchmen at the nonoperating mill. Shortly after World War II the Stoiber home was purchased for a few thousand dollars and torn down for its materials. Only the foundation, built from stone quarried nearby, remains today.

The foundation of the Silver Lake Mill is visible across the river, just a short distance upstream. The original Silver Lake Mill was located at the mine miles away and much higher near the south end of Arrastra Gulch. Shortly before the turn of the century this

The Shenandoah-Dives (Mayflower) Mill was connected by tram to its mine in Arrastra Gulch (behind the mill). Silver Lake was behind the snow-covered ridge and the Silver Lake Mill is to the far right in the photo. Courtesy Denver Public Library, Western History Dept. (X61018)

155

mill was built and connected to the mine by an eighty-four-hundred-foot tram (one span cleared twenty-two-hundred feet). The mill burned in 1906 but was rebuilt. It was purposely burned in the late 1940s for safety reasons.

Just past the interpretive sign, on the main road to the north, is the Mayflower Mill. Charles "Papa" Chase built the mill in 1933 to process ore for his Shenandoah-Dives Mine, which was a consolidation of many of the mines up Arrastra Gulch, including the Mayflower. The mine's offices, assayer's office, and machine shop were also at the mill. It operated for almost sixty years and at times had as many as forty mines shipping their ore. In the last ten years under Chase's management it was milling mainly low-grade ore, but since the mill was so important to the Silverton economy, Chase kept it open, even though it was losing money. It was reconstructed in 1960 to process ore from the Sunnyside Mine, which came out of the American Tunnel at Gladstone. Over 1.5 million ounces of gold and thirty million ounces of silver were processed at the mill, as well as many hundreds of thousands of pounds of base metals such as lead, zinc, and copper. When the Mayflower closed in 1991, it was the second longest running gold mill in the United States. In 1996 the Sunnyside Gold Company donated the mill to the San Juan County Historical Society and also gave them one hundred twenty thousand dollars to convert the mill to a tour site. It now serves as a wonderful mining museum—left basically as it was when it closed, but open only in the summer.

Across from the Mayflower Mill, a road dips down to the south, crosses the Animas River, and heads into Arrastra Gulch. Much of this road is four-wheel drive, and many of the mines in the gulch can be reached only on foot; but this is the area where mining began in the San Juans, and it contained some of the richest mines in the San Juans. There is evidence that the French might have been in Arrastra Gulch in the late 1790s, unauthorized Spanish were probably mining here as early as 1800, and Americans were prospecting the area in 1860. In 1870 the Little Giant was discovered, and the first milling equipment in the San Juans was brought to the mine and was operating by 1873.

Mining continued in Arrastra Gulch until the 1980s. It was here that some of the most complicated trams in the world were built, many of them using simply gravity for power, and it was here that Charles Wifley perfected new milling equipment and techniques later used throughout the world.

As you enter into Arrastra Gulch proper, the road to the Aspen Mine is about a quarter of a mile to the southwest on Hazelton Mountain across Arrastra Creek. In fact the road travels up the creek itself for a few hundred feet. The first right-hand fork on the Aspen road (a little more than a tenth of a mile) leads back to what was the upper part of the Silver Lake Mill. The remains of the tram to Silver Lake and the Aspen are still visible above the mill. Today, it makes a great picnic spot. As you continue to travel further up the side road to the Aspen Mine you will pass under the steel Mayflower tram towers. The wooden trams along the road went to the Silver Lake and the Aspen Mines. The roads to the right are private property. The first road to the left leads down to Arrastra Creek and out to the Whale Mill site. It should be traveled only on foot. The ten-stamp Whale Mill was erected in 1888 and was water-powered. At least a part of the mill might have come from the Little Giant operation. The Whale was later incorporated into the Silver Lake Mine operation.

At one mile from the creek, the upper road goes to the Aspen Mine dump and then a little further another road to the left leads to the upper workings of the mine. The road to the right leads to Aspen Camp, where the remains of four or five buildings are still standing. The building on the left was the boardinghouse. The early day camp of Quartzville was also near the site.

The Aspen Mine was an 1870s consolidation of many of the rich Hazelton Mountain mines, and it remained a good producer for many years thereafter. Tom Blair discovered the Aspen. Ingersoll noted "nothing but a bird or a mountain sheep would likely attempt the almost vertical wall rising from the southern side of Arrastra Gulch to culminate in the spires of Hazelton Mountain," so prospectors came in from this direction. The Aspen ore was originally milled at the Greene Smelter in Silverton. In 1879 Fossett reported the Aspen to have "produced more silver than any other mine in the San Juan country.... Its yield was $165,000." Its three- to eight-foot vein carried galena and gray copper and averaged about two hundred dollar per ton. Its claims included the Susquehana, Mammoth, Prospector, and Legal Tender—all big and early strikes. The Prospector Mine by 1876 had yielded sixty thousand dollars in ore. The Susquehanna produced forty thousand dollars' worth during the same time.

D&RG investors, who also bought the Greene Smelter and moved it to Durango, bought the Aspen. At the time of his visit in 1885 Ingersoll asked one miner about the weather and was told, "Its nine months of winter and three months of mighty late in the fall." But he also noted that the temperatures are warmer in the mountains than in the valley, that the men brought in their yearly supply of food and fuel during the summer, and that working underground the temperature was about the same all year (the high 50s and low 60s). Ingersoll's friend also pointed out that the locals liked to "snowshoe" on "Norwegian skidors—thin boards ten to twelve feet long, and slightly turned up in the front.... If you stay here in the winter you must learn, unless you are willing to be cooped up in your cabin from November till May." Ingersoll reported that since the men were basically stranded away from the "temptations of the barroom" that there was no other season of such great progress at the mines.

The road to the upper Aspen Mine tunnel is steep and has sharp rocks so most drivers will want to turn around at Aspen Camp. It also is usually possible (although hard) to continue across the mountain to the Lackawanna Mill, but at this point we will return to the Arrastra Gulch Road.

About another half mile up the main Arrastra Gulch Road there is an informal campground near the steel Mayflower tram towers, buckets, and cables. The Mayflower was the only mine in the area to use large steel tram towers instead of the small wooden ones commonly found throughout the San Juans. The towers were prefabricated, numbered, and hauled to their sites by mules in 1929. They were erected by hand. Trams were obviously important in Arrastra Gulch because of the harsh winters with deep snows and avalanches. It allowed the mines to greatly lower their shipping costs and to stay open year round-giving economic stability to the entire region.

About eight-tenths of a mile into Arrastra Gulch the roads become confusing. At the three-way fork it looks like the main road follows the creek to the right, but that road actually dead-ends into private property at the Iowa-Tiger mill site (in the basin) and the Ezra R. Mine (on the creek). The Ezra R. was located in 1882 and was worked until

91375 IOWA MILL - SILVERTON, COLO.

The Iowa Mill sat right in the middle of Arrastra Gulch. It was connected by tram to its mines in Silver Lake and connected by tram to its loading facilities on the Silverton Northern RR. Courtesy Denver Public Library, Western Hist. Dept. (Z2560)

fairly recently. The middle road also dead-ends into the Iowa Tiger. Keep to the left to travel down the Arrastra Gulch Road. The 150-ton Iowa-Tiger was built in 1900 to process ore coming from Silver Lake. The mill was connected by a 14,375-foot tram to the Iowa Mine, which in turn was connected by a tram to the Tiger. There was another tram from the mill down to a transfer station on the railroad siding by the Animas River. For two years, while the Mayflower Mill was being built, the Iowa-Tiger also processed Shenandoah-Dives ore. The Iowa Mill probably held the record for avalanche damage. It was hit eight times (twice, the destruction was total), and its tram was damaged four times.

Across the canyon to the south are the Grey Eagle and Oriental Mines, both located in the late 1890s. The basin up and to the right is Whale Basin. The Unity, Essex, Streak, and Whale Mines were located there, and the Unity tram came off the mountain and ran to the Silver Lake Mill.

The next road to the left leads to Little Giant Basin (which doesn't actually hold the Little Giant Mine). As the road switchbacks up the mountain it passes under the wooden tram of the Contention or Big Giant Mine several times. At a little less than two miles (at the small lake) are the ruins of the Black Prince Mill and Mine, which leased the claims of the Big Giant and Contention, which are located farther up in the basin. The Black Prince's owners attempted to drive a tunnel under those mines in 1915 but abandoned the project after it was only in 1500 feet. A small camp was built at the Black Prince site. The Contention tram and mill were used to process the ore. The Contention tram in turn was originally used by the North Star Mine but was later moved to the Contention Mill, and then moved downhill again to originate at the Black Prince. The Little Giant Mine is located downhill in the gulch to the southwest of the Black Prince.

The King Solomon and Jura Mines are on the left and uphill from the Black Prince, and about a half mile into the Big Giant Basin, the remains of the Big Giant or Contention Mill are to the west (right) and the North Star Mine is uphill and directly ahead (south) about a quarter of a mile and across Gold Lake. Even further up but not visible to the left in Dives Basin is the Shenandoah. The Big Giant was started in the 1870s but was never successful and was bought out by the Contention. The road dead-ends

The North Star clung to the very top of the mountain. Its miners climbed on top of its buildings for the photo. Courtesy Denver Public Library, Western History Dept. (X62714)

near this point (about three miles from its start) although it is possible to hike much further up into the mountains.

The mine ruins on the slopes straight ahead are the Number 5 Level of the North Star Mine. Its other levels are higher up and on the other (north) side of Little Giant Peak. The workings that are a little to the left and ahead are the McMillan Mine, which was on the Gold Lake vein (this was very close to where Edward Innis thought he would find a lake of gold). Because of its close proximity to Cunningham Pass, the North Star was located early (in 1872) but not officially staked until 1876 by Martin Van Buren Wason. Ernest Ingersoll in *The Crest of the Continent* pointed out "the ore is galena and gray copper of extraordinary grade. A marvelous trail has been cut through the woods and then nicked into the almost solid rock of the bald mountain-crest, far above timberline, or built out of logs, along which burros carry to the mine all its supplies, and bring back down its product." By 1877 the mine was shipping good ore-a little over one hundred ounces of silver per ton. However, work still went slowly—the miners were only in about opne hundred feet by 1879. That year the mine was taken over by the Crooke brothers, who had bought a partial interest in 1876. The North Star was later held to be the "highest consistent producing mine on the North American continent," and Crofutt declared it to be "the mother lode" of the claims on King Solomon Mountain"—which was saying a lot.

In 1879 the North Star and Royal Tiger were combined on the same vein, and in 1879 the mine was shipping ore that was 150 to 240 ounces of silver and fifty percent to sixty percent lead from a forty-foot wide vein that was three miles long. The Crookes also filed on a mill site and even a town site in the basin. There was a small sampling works with a tram to the mine from the basin. A wonderful account of a few years at the North Star Mine is contained in *San Juan Gold* by Duane Smith. Although never a very big mine (it employed only ten to twenty men), it did produce $1.3 million by 1897. At that point it began to fade into history, but the North Star was later mined by the Mayflower and Shenandoah-Dives.

Back on the Arrastra Gulch Road one can continue toward the Mayflower Mine. About a half mile further, as you come out of the trees, the Little Giant Mine is located on the left up French Gulch, later called the Little Giant Gulch. The Little Giant was one of the San Juan's first big mines. The first cabins in the San Juans were built around the mine. It was discovered in 1870 by Miles T. Johnson and originally crushed its ore in Spanish *arrastras* (hence the name for the gulch). It was abandoned by 1874 because of disputes among its owners, who even posted armed guards and drew guns on each other. Later it was again opened and worked until the 1970s. The Little Giant Mill was the first mill in the San Juans. It was hauled over Stony Pass by burros in 1872 and had five stamps. The mill was located one thousand feet below the mine and was connected by a rope tram. In 1872 a company was formed in Chicago to work the property, and investors put in enough capital to bring in a 12 horsepower engine, crusher, ball pulverizer, and amalgamation works. About $14,500 in ore was mined (at sixty-five percent of assayed value at the smelter). The Little Giant vein was described in 1875 as eight inches of gold-bearing quartz. Later, when worked by the Shenandoah-Dives, a small mill was built inside the mine in a worked-out stope.

As you travel up Arrastra Gulch, another steel Mayflower tram is to the right, and you can begin to recognize the many avalanche defense structures that were created up the mountain. About a half-mile further the remains of the Mayflower Mine can be seen scattered along the hillside to the left. After the switchback, the last steep section of road to the mine must usually be traveled on foot. Even if it has been recently maintained, there is a very sharp switchback in the middle that requires backing up in a steep and narrow section of road.

The Shenandoah Dives boarding house, offices and tram house at the Mayflower Tunnel clung to the side of the cliffs. Courtesy Denver Public Library, Western History Dept. (X62244)

The Mayflower was discovered in the late 1880s. It had a five-story boardinghouse and a small mill that clung to the hillside. It was consolidated with the Shenandoah-Dives in the 1920s. Mayflower production peaked in 1929 when it was connected by tram to its new mill by the Animas River. The tram operated from 1930 to 1953 and then from 1959 to 1961. By this time the mine encompassed (as you went into the mountain) the Mayflower, Slide, Terrible, North Star, Dives, and Shenandoah. The mine's boardinghouse contained offices, sleeping rooms, a commissary, billiards, a library, a dining room, and a kitchen. The stock market crashed just as construction was being completed, but the presence of gold and the grit of Charles Chase kept the mine afloat when other mines were closing. The Mayflower buildings were torn down in the 1960s.

The Silver Lake Mine's complex of buildings stood directly beside the lake where they dumped their tailings. Author's Collection.

John Marshall and Zeke Zanoni wonderfully describe life at the Shenandoah and other mines close to Silverton in *Mining the Hardrock.*

From the Mayflower Mine a trail continues on to Silver Lake. The hike is steep and sometimes sections of the trail are covered with snow, but it is well worth the effort. There were four main mines surrounding Silver Lake. The Iowa and the Tiger were across the lake from each other but operated together at times (a small tram ran across the lake to connect them). They were connected to the mill in Arrastra Gulch with a 14,375-foot tram. The Iowa and the Tiger were discovered in the 1880s.

The Silver Lake Mine was located in 1873 but not discovered to be profitable until 1890, when its mill was built at the lake. The Silver Lake also had a post office called Arrastra from 1895 to 1919, but there was no real town—just the post office. The mine was the second biggest producer in San Juan County in 1902. The mill at the lake could process four hundred tons of ore a day. The concentrated ore was then shipped down the 13,730-foot tram to be further refined at the mill by the Animas River. Between 1913 and 1919 the mine was leased to Otto Mears and Arthur Wilfley (who invented the Wilfley table, which processed gold, and a new process for gold recovery). The tailings from the mill by the lake were sent down a water flume to the mill below. The tailings from the Silver Lake were successfully remilled at their small mill located near Highway 110 and the Arrastra Gulch turnoff.

The ore at the Silver Lake Mine was relatively low-grade compared with other mines in the district (averaging fifty ounces of silver and sixty percent lead after concentrating), yet the Silver Lake Mine eventually worked over a hundred claims and produced over eleven million dollars from seventeen and a half miles of tunnels. The Silver Lake treated its miners well. Its four-story outhouse had the distinction of an opening for each floor. The mine in the gulch behind the lake is the Buckeye. Because

161

roads never connected to them, the Silver Lake mines remain better preserved than other comparable mines abandoned for that length of time. The mine is well worth the long, steep hike!

Back on State Highway 110, the Mayflower Mill is on the left; its ore cars still hang from the tram cable across the river. The tram was constructed in the early 1930s. If you look closely, about a half-mile further, you can see the Contention Mill site just before the swinging bridge. The mill was built in the 1890s to treat the ore brought down by tram from the North Star, Black Prince, and Big Giant Mines, which are located in Little and Big Giant Basins. The mill clung to the hillside right above the tracks, but it was destroyed by fire in the 1920s. The remains of a wooden flume can also be seen above the river. The flume was used to carry water to the power plant at Waldeim.

About a half mile further, and shortly before the road crosses the Animas River, the site of the original Pride of the West Mill can be seen on the road side of the river. It treated ore from its mine, which was discovered in Cunningham Gulch in 1871. The side road to the left, just before the bridge, is the Silverton Northern Railroad grade. It accesses several mines on the north side of the Animas River, but it is not always open to vehicles.

The area around Cunningham Creek, five miles east of Silverton, is the site of Howardsville (originally called *Bullion City* for a short while). Bullion City was only an attempt in 1874 to capitalize on a town filing, and there is no evidence that its founders ever followed through. For a few years the area was called by either name, but by late 1875 the name Howardsville had been settled on. The first store and saloon were built by Henry Gill and G. S. Flagler. Howardsville was the first settlement on the Western Slope of Colorado, and it became the first county seat of La Plata County, which was much bigger than today—composed of what is now La Plata, San Juan, Ouray, Montezuma, Dolores, and San Miguel Counties. Howardsville was named for George Howard, who had prospected the area with the Charles Baker party in 1860. It was said that Howard got his cabin built with the help of a lot of whiskey. "Passers-by were offered a drink of whiskey as pay for their assistance in lifting his logs into place. He had no trouble getting assistance."

Howardsville was touted as the only place in the San Juans where a man could buy two beers for twenty-five cents. This gave it a real head start as the early social center for the Animas Valley. (Fischer's Rocky Mountain Brewery was also one of its first commercial establishments.) On June 28, 1874, a post office was established—but no actual delivery occurred until early 1875. It was still the first post office on the Western Slope. In addition to the brewery, Howardsville boasted a hotel, two restaurants, several saloons, and many cabins. The first white child (a boy) delivered in the San Juans was born in 1874 to Carrie and William H. Nichols, residents of Howardsville. Mr. Nichols was also the local postmaster. The Wilson Division of the Hayden Survey had its main camp in 1874 and 1875 in the little meadow on the bench to the southeast above the Howardsville Mill. The Howardsville cemetery is now near the spot. By the end of 1874 Howardsville was twice the size of Silverton. In 1875 Howardsville had a summer population of almost three hundred, but Silverton had grown quicker. That same year, by a vote of 183 out of 341, Silverton was made the new county seat.

This is an early photo of Howardsville by William H. Jackson as most of the houses are log cabins. Author's Collection.

An interesting story is told of an event that occurred in Howardsville after the Meeker Massacre in 1879. A man named McCann rushed into a local saloon screaming that the Utes had massacred everyone in Animas City and were headed this way and then asked for a drink. He said he was headed across Stony Pass to Antelope Park to warn others there so the bartender gave him some fortification. Later, after a massive panic, it was learned that the man had made the whole affair up just to get a free drink.

The arrival of the D&RG in Silverton in 1882 (where it ended) pretty much doomed Howardsville from developing into a good size town. By 1885 Ingersoll noted it "was the center of everything a few years ago." Crofutt in 1885 wrote "the village proper consists of several stores and saloons, one small reduction works, about 30 buildings of all kinds and a population of about 100." By 1888 Howardsville's population was down to less than fifty, as most of its citizens had moved to either Silverton or Eureka. The town rebounded in 1894 when it became an important stop on the Silverton Northern Railroad, but by the 1920s it was again in decline. In 1937 the branch railroad up nearby Cunningham Gulch was abandoned, the mines shut down, and more people moved away. The post office was discontinued in 1939. Very few of its original buildings exist today. A small mill, built in 1921 for the Little Nation Mine, still has tramlines in place, running to the various adits on the north side of King Solomon Mountain. The red home on the bench was built in 1874 by Thomas Trippe, a surveyor. The metal building across Cunningham Creek is the second Pride of the West Mill, also known as the Howardsville Mill. It was built in the 1940s and worked into the 1980s. The building with the four-sided roof is the old Howardsville School.

Howardsville was built at the mouth of Cunningham Gulch (named after W. H. Cunningham, one of the original Charles Baker's party in 1860). Many of the original

San Juan prospectors came from Del Norte over Stony Pass or Cunningham Pass and down Cunningham Creek to the Animas River. It was an extremely dangerous trip in the winter, because Cunningham Gulch is one of the most avalanche prone areas in the world, yet the gulch was the site of many successful mines and mills, some of which operated year round right up to the present. The gulch is composed of King Solomon Mountain to the south and Galena and Green Mountains to the north. The biggest mining operations included the Green Mountain, Pride of the West, Buffalo Boy, and Gary Owen. A short distance up Cunningham Gulch from Howardsville, the road forks. The two roads then parallel each other for the next mile or so. The upper road goes to the Old One Hundred Mine, and the bottom follows the creek. The first dump on the other side of the creek is the Little Nation Mine. Then about a mile up the gulch the road passes through the Old One Hundred Mine.

The Old One Hundred Mine was located by the Neigold brothers in 1875. The mine started slowly, but by the early 1900s was doing well. It was sold to the Old One Hundred Mining Co. in 1904 for four hundred thousand dollars. The mine was named for a hymn—maybe because the prospectors were probably praying by the time they got to the mine. Its vein was up to fourteen feet wide and ran straight up the mountain. This meant the mine was developed on different levels, but because the ore at lower levels was a much lower grade, the first level was almost one thousand feet off the valley floor. Unfortunately, the rich ore was found only in small pockets.

The Old One Hundred Company covered thirty claims and encompassed a good part of the mountain. It eventually had over three miles of underground workings. The mine and mill were connected by a 3,700 foot tram. It is now perhaps most famous for its three-story boardinghouse and tram house, which cling to the cliffs near the top of the mountain above the mine. The mine had boardinghouses at several levels on the mountain, but this one is the most famous. The mill was built in 1905, had forty stamps, and could crush two hundred tons a day. The mill worked its way down the hillside with its power plant located at the bottom. The Old One Hundred did well—sometimes recovering forty thousand dollars in gold a month (the equivalent of a million dollars today). It was supposed to have sent a gold brick to the Denver Mint in 1906 that weighed fifty pounds and was worth twelve thousand dollars (three hundred fifty thousand dollars in today's money). Most gold bars were just half this size—the mine owners were just looking for publicity. The Old One Hundred Mining Co. failed in 1909, and the mine wasn't worked from 1909 to 1934. It opened again for a few more years, then closed again. Dixilyn Corp. worked it in the 1960s and 1970s. Today, a very interesting mine tour operates during the summer and affords you a chance, for a reasonable fee, to explore an actual mine and see mining equipment in action.

The Neigold brothers arrived in 1872, prospected all up and down Cunningham Gulch, and built a cabin in 1873 at the intersection of Cunningham and Stony Creeks. The four brothers each had their own cabins by 1876, and they founded Niegoldstown, an early settlement with a post office from 1879 to 1881, a boardinghouse, store, and a few cabins. The Neigolds came from a well-to-do European family. They lived a good life in their little town, and they even talked of building a stone Saxon castle in the gulch. They were supposedly backed by a rich uncle in Philadelphia. They shipped in a grand piano, sang opera for their friends, and imported fine wines and tobacco. They built

The Old One Hundred Mill is on the left and Stony Pass near the middle of this photo of Cunningham Gulch taken about 1890. Author's Collection.

the Tecumseh Mill on the site, mainly to serve the Pride of the West and Philadelphia Mines. The Philadelphia was owned by the Niegolds and had ore that ran up to $1,128 in silver per ton. By the winter of 1875-76 its ore was down to averaging two hundred to eight hundred dollars per ton at the Greene Smelter. The rich ore turned out to be in pockets but it produced enough money for the Neigolds to continue their development work. Besides their five-stamp mill they had a sawmill. They kept about twenty men working on various claims and supposedly sent a shipment of ore that was about the size of a boxcar to Germany, which netted a hundred thousand dollars in ore. It was a successful settlement until 1884, when the mill and three dwellings were destroyed by a snowslide and were never rebuilt.

In 1905 a spur from the Silverton Northern Railroad was run up Cunningham Gulch all the way to the Green Mountain Mill, which is a little less than a half-mile further up from Neigoldston and downhill from the road. The Green Mountain Mill and a new tram were built in 1904, and it treated ore from mines that were located on Green Mountain further up the gulch. It was connected by its three thousand-foot tram to the Osceola (Green Mountain) Mine. It was hit by a snowslide in 1906, just weeks after it was completed, but only one man was killed because the other workmen had gone to town, and it was soon rebuilt.

As previously mentioned, Cunningham Gulch is one of the most avalanche-prone areas in the world. 1906 was an especially bad year. On March 17 a slide hit the Green Mountain Mill, doing fifty thousand dollars in damage and the next day the Shenandoah boardinghouse (located in the basin above Cunningham) was hit. Of the twenty-one miners in the building, only nine survived. The March 24 edition of the *Silverton Standard* reported "almost four miles of its length is an unbroken series of slides and in places the gulch is buried under snow 150 feet deep."

Shortly past the Green Mountain Mill is the Buffalo Boy Tram House (it was not the original tram house)—constructed in 1936. At one time there was a mill and boardinghouse at the site, but they burned. The Buffalo Boy Mine was discovered in 1883, and the original tram was built in 1928 for a distance of 8,740 feet from the mill to the mine, to a spot high up on the Stony Pass road. Its miners usually rode the tram to work.

At a point about a quarter mile further, the upper and lower roads rejoin and shortly thereafter the road forks. The right-hand fork continues up Cunningham Gulch and dead-ends shortly above the Highland Mary Mine. The left fork takes you over Stony Pass and the Continental Divide to the drainage of the Rio Grande. Stony Pass became the accepted route for travel into Silverton from 1870 to 1882 when the D&RG Railroad arrived in Silverton. Stony Pass has also been used to designate both the present-day pass and Cunningham Pass at the very end of Cunningham Gulch. The original Ute trail was over Cunningham Pass (which is five hundred feet lower than Stony). The Cunningham route was also used by the Baker party in 1860 and undoubtedly was used by countless unknown Spanish prospectors.

The first "road" down Stony Pass was "built" by Major E. M. Hamilton in 1872 to transport the Little Giant Mill machinery. Stony Pass was called *Hamilton Pass* for a while after that event. However, Hamilton's men were doing little more than throwing rocks out of the way of the wagon wheels. Freighters claimed that the early trail was so rough that "every ten feet there is a stone projecting from six to eighteen inches out of the ground, and frequently on the opposite side of the trail a hole from six to eighteen inches deep with a stump in the middle." The "road" was reported by one man to have been the "author of more profanity than any other evil we endure." The pass, although high, was usually open by early May compared with other routes (including Cunningham Pass) that might not be open until late June. When Ruffner used the passes in 1873 he tried out Minnie, Maggie, Stony, and Cunningham, and he suggested Cunningham as the best route for a wagon road. He wrote, "Hamilton's route could not even be located down Stony." The Wheeler Survey in 1873 through 1875 decribed the average slope on Stony to be twenty degrees and stated it "is not practicable for wagons, although there have been many wagons over it." In 1874 Ernest Ingersoll, who was traveling with the Hayden Survey, noted that (Stony Pass) "is called a wagon road; but the only means of using it is to take your wagon to pieces and let it down several steep places by rope." Most people avoided the problem by leaving their wagons on the Rio Grande side and coming the rest of the way on horseback or on foot.

The story is told that Mrs. George Webb, coming over Stony Pass on June 5, 1875, delivered her baby just to the north of the summit in a shelter made of the wagon's cover and on a bed of spruce branches spread on the snow. "Timberline Webb" was the first female child in San Juan County, and the citizens of Silverton presented a city lot to the girl that came to be known as "the snow baby." On a sadder note, the mail carrier (also the sheriff) John Greenell, froze to death in November 1876 while trying to get the mail over the Stony Pass. The paper reported, "If anyone in San Juan County could be said to have everyone for his friend, and no one for his enemy, he surely was that man."

The 110-mile route from Del Norte was the shortest way to Silverton, and most important before 1881, it did not cross through Ute land. For the first ninety miles

The Pride of the West, the first big mine in Cunningham Gulch, was still doing well in 1937 when this photo was taken. Courtesy Denver Public Library, Western History Dept. (X62203)

This is all that was left of the Green Mountain Mill after it was hit by an avalanche on March 17, 1906. Author's Collection.

the route was very easy and somewhat inhabited until the traveler reached Lost Trail Creek on the eastern side. On either Stony or Cunningham Pass, the part of the route from the top down to Cunningham Creek was by far the hardest—a drop of fifteen hundred feet in two miles. By 1879 an actual wagon road was made over Stony Pass, so traffic shifted from Cunningham Pass to Stony. Regardless of which pass was taken, Ingersoll noted, "The whole town would be alive with a general jubilation when the tinkling of bells of the first train of jacks was heard in the spring, for that meant the end of a six-month siege in the midst of impassable snow." It is hard to imagine the amount a goods and equipment brought into the San Juans over these very rough roads between 1875 and 1882.

At first, freighters traveling the Stony Pass road cut off small trees to drag behind their wagons, but later W. D. "Squire" Watson set up a way to lower the wagons by using ropes and large stumps. Some say the fee was $2.50 for a 150-foot trip, but it was more likely that Watson used several more trees over a longer stretch to gain his money. It was a highly profitable venture for on a good day as many as twenty wagons might pass over the road. In 1910 the first car came over Stony Pass (although it had to be pulled a good part of the way by horses). It was a French Croxton Keetof (30 horsepower) and caused quite a commotion! People started lining the road at the top of the pass, and children ran alongside it just trying to get a chance to touch it.

Today, as we travel the Stony Pass route we almost immediately encounter an impassable road leading to the right that went to the Osceola Mine. When the Stony Pass Road crosses the creek you can get a good view of the older, steeper route—not recommended any longer! You might think the current road is bad but think of what it was like originally. About a mile and a half up is a private road, but the next road to the left (San Juan County 3B and about a quarter of a mile further) leads to the Gary Owen Mine. It was worked by the Old One Hundred Mine from the 1940s to 1960s. There is also a foot trail that leaves uphill from this road to the Buffalo Boy Mine. Above tree line on Stony and directly below the road to the right is the Omaya Tunnel—driven by Homestake Mining Co. in the 1980s to access the Buffalo Boy (parts of which can be seen from the north). The top of Stony Pass is a little over four miles from the Cunningham Gulch Road. The elevation is 12,650 feet.

The other side of Stony Pass (which is on the eastern side of the Continental Divide and contains the headwaters of the Rio Grande) is less steep but there are some bad spots. Pole Creek (usually good fishing) is about six miles down from the summit and might be impassable during the spring runoff. The Weminuche Wilderness area is just to the west of the road on the way down. Because it is a long way to a town of any size, let's turn around and return to Cunningham Gulch.

Continuing on the road up Cunningham Gulch (San Juan County 4A), the Lawrence Tunnel can be seen down by the creek and the Osceola Mine above the road. The Osceola was discovered in 1877, and the Lawrence Tunnel was driven to intersect it at a lower level in 1900. A tram then carried the ore down to the Green Mountain Mill. A little further down the road the Pride of the West Mine can be seen on the left (east) side of the road. It was discovered in 1871 and was the first of the San Juan mines to apply for a patent. In 1874 it shipped ore worth $60 per ton over Stony Pass. The Pride of the West had ten men working in 1878, and its ore was up to a hundred dollars per ton. Its first mill was erected about 1881. For a long time the Pride of the West looked like it would be the best mine in the area because it had a seven-foot wide vein of solid rich galena. It was by far the biggest producer of the mines in Cunningham Gulch and has very extensive workings. In 1900 it employed one hundred men. Its mills were in Howardsville. In 1948 eight miners were killed when smoke from a fire outside the portal was pulled into the mine. It had a two thousand-foot tram to the Silverton Northern at one time. The Pride of the West produced over six million dollars in ore.

A little further on (about three miles from Howardsville) is the Green Mountain Mine. It was discovered in 1874 by Dempsey Reese, Bill Mulholland, Mickey Breene, and Tom Blair and was worked off and on until 1950. In 1904 it was connected by tram to its three hundred ton-mill back down the valley. Its claims include the Leopard, King William, and the Flat Broke. The Green Mountain vein was similar to the Pride of the West, but not as big. In 1883 its ore ran thirty-eight ounces of silver and forty-five percent lead. However, its concentrated gray copper ore assayed at eighty to five hundred ounces of silver per ton.

In another quarter of a mile there are sheep pens and also a public restroom. The power line running to the Shenandoah-Dives Mine is also visible on the west side of the gulch near the top of the mountains. The mine was an early discovery but was not worked much at first. In the winter the miners were reported to get up and down to the mine on skis. In March of 1906 a massive avalanche killed twelve men while they were eating supper in the boardinghouse. The grotesquely twisted bodies were taken to the morgue to be thawed out and straightened up before relatives were allowed to see them. The Shenandoah did well for its owners. In the early 1890s two pack trains a day brought the rich ore down to be milled. After the Silverton Northern arrived, the Shenandoah and Highland Mary ore was shipped to the Green Mountain Mill where the railroad spur ended.

The Highland Mary Mill is about a half mile further, where the road crosses the creek. It was located in 1874 by John Dunn, Andrew Richardson, and William Quinn. The mine was 1,000 feet up on the cliffs to the south. Its original mill was partially made up of equipment from the Little Giant. In 1875 Edward Innis bought the mine for thirty thousand dollars and used a New York spiritualist (who wrote letters of instruction) to drive a three thousand-foot tunnel. She was eventually paid fifty thousand dollars and told him that a "lake of gold" was to be found within two thousand feet of where he started the tunnel. Innis started work with thirty men on the Innis Tunnel (located by the creek) and good ore was found at the outset-some small pockets even assayed at seven thousand dollars per ton. However, the miners were often spooked by the use of a spiritualist and refused to enter the mine alone. The tunnels twisted and turned in every direction and actually hit a few silver veins, but not much gold was discovered and the tunnel was abandoned in 1884 after almost a million dollars had been spent on its development. Innis died penniless in an insane asylum in 1900.

Like most of the nearby mines, the Highland Mary had a major avalanche disaster. On February 12, 1879, seven of nine miners made heard the noise and made it through a trap door in the boardinghouse floor and into the tunnel at the Number 4 level, but two miners were killed and later two other men were killed while attempting a rescue. The Highland Mary was considered one of the great early mines of the San Juans. Eventually,

The Highland Mary Mill about 1910. The Innis Tunnel was just to the right of the mill. Courtesy Denver Public Library, Western History Dept. (X62209)

Innis built thirteen buildings on the site, including stables, a dining hall, offices, an engine house, and his two-story home, which was known as "the White House" and which had a post office from 1878 to 1885. It was located by the first switchback past the mill, but burned in 1942. The White House was for many years an important rest stop for those traveling over Cunningham Pass.

Crofutt in 1885 identified the Highland Mary as a "small mining town." He also felt that all the important Silverton mines were located in Cunningham Gulch. "The Highland Mary Mine, that runs $300 to $500, is the chief among a score or more that approximate it closely. The ores are an argentiferous galena with gray copper of high grade, some select ores running as high as $5,000 per ton." In 1902 the Highland Mary built a fifty-ton concentrating mill. In 1907 the mine's output was the second largest in San Juan County. The mine was worked until 1952 and made about three million dollars.

The road continues for about a mile past the mill, where there is a cutoff to the Highland Mary Lakes Trailhead—a great hike and usually

The Sunnyside Mine had many buildings high above timberline surrounding Lake Emma. Author's photo.

The "White House" was an important stop for travelers going over Cunningham Pass, even in 1906. Courtesy Denver Public Library, Western History Dept. (X21782)

decent fishing. This trail can be followed all the way over the Continental Divide. The main road continues for a short distance, but it is advisable to turn around here. The trail at the end of the road goes to Spencer Basin alongside Mountaineer Creek. The Mountaineer Mine along this path was one of the very first mining claims filed in the San Juans.

Back on the main road from Silverton to Animas Forks (County Road 110), the Buffalo Boy Mill is on the right and the Howardsville tailings dump starts soon thereafter on the left. The Silverton Northern Railroad grade is visible across the Animas River for many miles. After you pass the beaver ponds on the left, the remains of the Hamlet Mill and its 1800-foot tram are on the right (about a mile out of Howardsville). It was built in 1875 and destroyed by fire in 1915.

Just past the Hamlet Mill is Maggie Gulch, which is a four-wheel-drive road (San Juan County 23). There are also restroom facilities at this intersection. The "town" of Middleton was started about 1881. T. M. Crawford built one of the first cabins, so it was called Crawfordville for the first couple of years. Middleton covered the area from the mouth of Maggie Gulch all the way up to Minnie Gulch. It had a school, post office, and several houses; but it was already a ghost town by the early 1900s. People chose

The Hamlet Mill was still in place in 1942 when Muriel Wolle took this photograph. Courtesy Denver Public Library, Western History Dept. (X61025)

to live in nearby Eureka or Howardsville instead. The town got its name from Middle Mountain, which separates the two gulches and which in turn was named for being halfway between Howardsville and Eureka. It existed mainly because of the nearby Hamlet and Gold Nugget Mines. The mines in Maggie Gulch were discovered later than the other mines nearby. Its real rush occurred in 1894 and 1895. Later Middleton had two mills—the Kittemac (closer to Minnie Gulch) and the Hamlet, both connected by trams to serve the local mines.

There are several mines up Maggie Gulch. The first is the Ruby (on the right by the creek after 1.2 miles). The Ruby produced large amounts of tungsten. The vein can be seen clearly above the mine. The large crevice is actually part of a stope that collapsed. Then the tram to the Ridgeway Mine (which was first worked in 1897) can be seen a short way further along the road above the waterfall. The mine was high above to the south on Galena Mountain. Then the Dewey, Gold Nugget, and the Little Maude mines are all along the road to the left. They are more visible when you're traveling back down the gulch. These mines were all worked about 1890 to 1900. About a mile further, to the right across the creek, is the Empire Mine (with its modern building) and finally at about four and a half miles from Howardsville are the Intersection Mill and Mine. The mill (including all of its ten stamps) is still in fairly good shape. Beyond the Intersection is a hiking and horseback trail, which crosses the Continental Divide to Pole Creek in the Rio Grande drainage. Early prospectors sometimes used this route instead of the better-known Stony Pass.

Muriel Wolle also took this photo in 1942 of the Kittimac Mill. Courtesy Denver Public Library, Western History Dept. (X61024)

After traveling back to the Silverton/Animas Forks Road (County Road 110) and proceeding about a half mile, the next road you encounter is Minnie Gulch. Minnie Gulch also contains many ruins and is a beautiful drive. In 1874 Minnie Gulch's veins were described as "frequent and many of them are of immense size but assay low (from six to forty ounces of silver per ton)." A toll road was proposed up Minnie and over the Continental Divide in 1876. The route was surveyed but never built. A little more than a mile up Minnie on San Juan County 24 is the Caledonia Mill. It was also connected by a tram to its mine. The wonderful stone foundation for its mill, the boardinghouse, and the mine superintendent's house are all at least partially standing. A little further, when the road forks, the route to the left goes to the Caledonia Mine (at about a mile) and then the Kittimac Mine and Mill a

The large Sunnyside Mill dominated the town of Eureka, which was at its peak in the 1920s and 30s. Author's Collection.

half mile further. The Kittimac was connected to its mill in Middleton by tram. These were early claims—discovered about 1872 but not really developed until about 1900. In 1902 Kittimac ore ran one hundred ounces of silver and 1.5 ounces of gold. Back on the main Minnie Gulch road the road to the right leads about three miles to the Esmerelda, which produced heavily about 1900. At the end of the road there is another pack trail over the Continental Divide.

Northeast of Minnie Gulch, towards Animas Forks, the main highway enters a wide portion of the valley where Eureka Gulch (to the north) joins the Animas River. This is the site of the town of Eureka (Greek for "I have found it). The extremely rich Sunnyside Mine was largely responsible for the town's existence, as it was in Eureka that the large Sunnyside mills were eventually built. The mine itself was three miles up Eureka Gulch and two thousand feet higher in elevation than the town.

Reuben J. McNutt probably built the first cabin near Eureka in 1872, but it was not within the town site but rather up Eureka Gulch. The first house in what came to be the city limits was built later that same year, and by 1873 there were several other cabins in the area. The town site was filed in 1874. The 1874 Hayden Survey party reported coming into "a thick clump of trees in which were several log cabins bearing on a flaring signboard the word *Eureka,* evidently intended for the name of a town that was expected to be...." Its post office was established on August 9, 1875, and was in continuous operation until May of 1942. In 1877 the town started growing quickly and by the end of that year had about thirty cabins and a reduction plant, which unfortunately never

173

did well and was used as a barn during much of its existence. Eureka was incorporated in 1883 and had several churches, a school, and sizable residential and commercial districts. However, Eureka seemed to have only promise and not a proven economy. In 1883 Burchard reported:

> Very little ore has been shipped from this part of San Juan County, because the mines as a general rule are above timberline and are somewhat inaccessible, and the ore has to be packed out quite a distance, then by wagons to Silverton for shipment to outside smelters.

Ingersoll reported in 1885 that Eureka was "a neat little village nestling among the trees. Here too are concentrating works, and the headquarters of several companies operating in Eureka, Minnie, and Maggie Gulches." Crofutt stated in 1885, "The town consists of one store, hotel, a dozen buildings, one smelting works, and a population of nearly 200.... Some of the best property at this place is locked up in litigation, which is a certain guarantee that it is rich in minerals."

The town of Eureka really boomed when the Silverton Northern arrived in 1897. In 1899 John Terry built a three-mile tram to Eureka to connect the Sunnyside Mine with a new mill (the smaller of the two Sunnyside mills eventually built in Eureka). In 1902 the Sunnyside Mine was third in production in San Juan County, and there were about four hundred men working at either the mine or the mill. Eureka's population peaked at about three hundred near this time. In 1918 an even bigger mill with a capacity of six hundred tons a day was built right next to the old mill at Eureka. The new mill included the first lead-zinc selective flotation plant in North America. It was later expanded to process twelve hundred tons a day. Much of the steel and timber for the mill came from the old Gold Prince Mill in Animas Forks. The first Eureka mill (which was wooden) burned in the 1920s. The mine and the mill were worked off and on until the early 1940s. Total production was about fifty million dollars. Many of the town's buildings were sold at a bankruptcy sale in 1948 because the mine owned about seventy-five percent of the town at that time. The houses were moved as far away as Silverton, the Idarado Mine, and even Ouray.

Very little remains of Eureka except the restored square water tower (which was first used by the railroad and then by the town for its water system) and the foundations and cellars of the town's houses. The enclosed area below the water tank was used to hold the town's fire cart and also served as the town's jail. One of the town's original cabins stands beyond the water tank but the

The Eureka school in 1904 was small but a very pretty building. Courtesy Denver Public Library, Western History Collection. (X5609)

floods of the Animas River covered up most of the evidence of the town itself. The massive foundation on the hillside to the west is what remains of the second Sunnyside Mill. The smaller first mill was to the south (left). There are good camping facilities in the Eureka area.

The road up Eureka Gulch (San Juan County 25, which is a rather easy four-wheel-drive road) is met just slightly uphill of the ruins of the mills and leads about three and a half miles to the former sites of the Sunnyside Mine and Lake Emma. There were two trams running from Eureka. One ran up to the mine and the other to the Midway Mill. The tram to the mine was 15,600 feet long. The dump above the mills is not connected with the Sunnyside but rather is from the lower workings of the Ransome Mine. Watch for the tram towers on the right as you continue up the road. At about a mile the road forks. The right road goes on up to upper workings of the Ransome Mine.

Then just a short distance further along the main road, a road goes to the left and down to the South Fork of Eureka Creek. On the way it passes the Terry Tunnel, which was started by John Terry in 1906. The Terry Tunnel was built to work the Sunnyside at a lower level, as the previous mining was very shallow and goes about a mile into the mountain. Five workers were killed while shoveling snow from the entrance to the Terry Tunnel in 1906. It was not a big slide, reported as only the size of a small room, but all five men suffocated before they could be rescued. The same winter the Sunnyside lost six of its tram towers to avalanches. The side road ends after about a half mile at the Midway Mill. The ruins of the Midway Mill are in a beautiful little valley with good primitive camping. A wonderful hike can be made up the South Fork from here. A lot of information about Eureka and the Sunnyside Mine can be found in Muriel Wolle's *Timberline Tailings*, a companion book to her *Stampede to Timberline*.

Terry had a home at the Midway Mill that has now collapsed. It was built in 1890, and he lived there during the summer months until about 1900, when he built a new house in Eureka. The Midway Mill was built in 1890. Its fifteen-stamp mill was powered by water from Eureka Creek. It had a thirty-ton capacity and switched to milling ore from the Sound Democrat Mine (located over the ridge) in 1900 when the new Eureka mill opened. There was also a sawmill on the site. The tram ran through the mill, which partially processed some of the ore. It also allowed ore buckets to be transferred from the upper tram to the lower tram.

The main Eureka Gulch road follows the second tram to the mine, and about a mile above the Terry Tunnel, the Sunnyside trams are very visible on the hillside to the left. The large structure is the tension station—necessary for such a long span of tramway. The tailings through which the road then passes are the Ben Franklin Mine, which was sold by George Howard to W. E. Webb and W. H. Whiton for thirty thousand dollars in 1882. Some of the mine was worked from the surface, and some of the shallow workings have collapsed. The nearby George Washington was owned by Edward Innis and later became an important part of the Sunnyside.

Then about three quarters of a mile further is what is left of Lake Emma (elevation 12,240 feet) and the Sunnyside Mine. Both have virtually disappeared as a result of reclamation efforts after the lake collapsed into the mine, but the mill was the first structure along the road, then the boardinghouses, and finally the mine itself.

George Howard (who founded Howardsville) and Rueben J. McNutt filed the original Sunnyside claim in 1873 when it was still Ute territory, so they refiled their claims in 1874. The initial discoveries were mainly silver. The two partners were constantly disagreeing on how to develop the mine, and they eventually became bitter enemies. Both men sold parts of their claims to others, and eventually the fractionalized ownership prevented development of the mine. The Thompson brothers worked the mine in the late 1870s and 1880s. Some years they were successful, mining as much as $385,000 in a year. By the late 1880s there was a self-contained community built around Lake Emma. The buildings were connected by covered snowsheds because the snow could pile up to twenty feet deep. In 1880 there were about two hundred men living around the mine. In 1882 the Sunnyside first struck gold in sizeable quantities. A bunkhouse and blacksmith shop were built at the mine that year. Some ore averaged nine ounces of gold and four hundred ounces of silver. They built a stamp mill at the portal but the ore values dropped, their costs rose, and the ore got harder to find.

In 1888 Judge John Terry built a ten-stamp mill at the mine for the Thompson brothers at Lake Emma (at the brothers' insistence), but it was hampered by lack of water and the high cost of coal. Terry was given partial ownership for his assistance. In 1894 Terry got control of the mine and built the Midway Mill. Terry kept buying partial interests in the mine until he had one hundred percent ownership. Terry spent so much buying ownership of the mine that just a short time later he was broke and was forced

The snowshed on the Silverton Northern Railroad at "The Big Slide" is slightly above Otto Mears' Toll road. The bridge's foundation can still be seen. Courtesy Denver Public Library, Western History Dept. Photo by George L. Beam. (GB8073)

to sell the mine to a New York syndicate that paid seventy-five thousand dollars down. However the New York group defaulted on the balance, and Terry used the seventy-five thousand down payment to continue work. He drove a crosscut into an extremely barren part of the mine, simply because he couldn't believe there could be such a large zone with minerals. The decision had been made to again close the mine when he hit a bonanza of gold in the Fourth of July Stope. If his randomly placed tunnel had been located a few feet either way, he would have missed the vein altogether.

By 1904 the mine had 180 men working. Covered walkways were used between the various buildings because of the deep snows at the mine's high location. In 1918 the new owner, U.S. Smelting and Refining Co., built the big new mill in Eureka. It had a one thousand-ton a day capacity. In 1919 a large fire destroyed most of the buildings at the mine, but they were quickly rebuilt. In 1927 the Sunnyside became the first Colorado mine to produce one thousand tons of ore in a day and employed over five hundred men. It continued to work sporadically until it closed in 1938, a fatality of the Great Depression. The mine remained closed from 1938 to 1959 when it was sold to Standard Metals and the access tunnel was moved to Gladstone. In 1976 a new mill was built in Gladstone. The new tunnel struck the vein six hundred feet below the old workings, and there were rich new veins discovered along the way. For a few years the Sunnyside was the richest gold mine in all of Colorado. In June of 1978, when the mine was being actively worked for gold, a stope in the mine caved in, drained Lake Emma, and left more than a million tons of mud in the mine. Fortunately the cave-in occurred on a Sunday afternoon when the mine was deserted or over one hundred men would have been killed. It took two years to clean up the mess and reopen the mine. After that, the Sunnyside was worked by Standard Metals in connection with other mines until 1985. Echo Boy operated it on a smaller scale until 1991, but there has been no mining since that time. The story of the Sunnyside Mine is well told in *Silverton Gold* by Allan G. Bird.

Back on the main road to Animas Forks, the highway is basically following the old Silverton Northern rail bed. This section of the railroad was completed in 1904. A good deal of the track was laid on Otto Mears's toll road. The engine had to push the cars from this point because the couplings often failed on its seven percent grade and hand brakes would not have held the cars. Service was discontinued in 1916 in part because the grade allowed a locomotive to carry only two loaded cars, which made railroad service very unprofitable.

After you drive just a short distance out of Eureka, the large boardinghouse to the right is the Martin or Tom Moore boarding house. The Tom Moore Mine, which owned the boardinghouse, was just a little further up the creek. The mine was located in the 1870s, and Samuel G. Martin built the boardinghouse in 1907. It is one of the best examples left standing of a large mining boardinghouse, but it is now used as a private residence.

About a quarter of a mile past the Tom Moore Mine, the old timbers to the left are what remains of a snow shed meant to protect the Silverton Northern Railroad from "The Big Slide." The snowshed was built into the mountain's slope and was made of extremely heavy timbers, bolted together and reinforced with a rock foundation. However, it was badly damaged during very its first year, and Otto Mears gave up on his idea of a series of such structures at each major slide between Eureka and Animas Forks. Thereafter, the road was often closed in the winter. The old roadbed and bridge supports that can be seen across the canyon are what remain of a bridge on the 1873 Otto Mears toll road from Lake City (it crossed to the east side of the canyon at this point). The toll road can then be seen clearly in the canyon passing through the mines and mills on the other side of the river.

The Silver Wing Mine and Mill is a little over a half a mile further. It was one of the great early mines of the area. Charles Jones, a member of the original Baker party, discovered the mine in 1874, and in 1875 the mine was described as ten lodes with two tunnels (one already in one thousand feet). Assays were running $130 to $2,000 per ton in iron, lead, copper, and silver. The 150-ton mill was built by the river in the 1890s and operated until the 1950s. Some work has been done at the mine right up to the present, such as the replacement of the bridge across the canyon. Snowslides were a major factor for the Silver Wing Mine. The Silver Wing Slide hit the bunkhouse killing one miner and injuring others. When two brothers of the dead miner came to retrieve the body, they had to abandon it in a terrible storm and then found it had been swept away by another avalanche during the night. They found the body only after considerable digging.

Next along the road are the Toltec Mine and Mill. The mine was discovered in the 1890s and worked off and on until present. In 1903 three men, Percy Kemper, Edward Crane, and L. W. Lofgren, were killed while thawing powder (believe it or not, this was done all the time to make the dynamite explode better). The explosion blew them to bits, and the small pieces of bodies that were found were placed in nail kegs. The cable bridge has obviously not been used in modern times and was meant only for men and horses—not loaded wagons. The mill was down by the creek, and there was a large boardinghouse on the east bank of the Animas River. The mine was higher up the cliffs on the west bank.

A little less than a half-mile further is the Picayune Gulch road, which departs to the left. It leads to Treasure Mountain and over to Placer (Mastodon) Gulch (discussed in another chapter).

Further along the Silverton/Animas Forks Road (Colorado 11) you can ford the creek at this point or go up the main road about two hundred feet to the bridge, cross over the river, and turn back to the right to Burns Gulch. This side road is rough and steep. It first goes by the Toltec Mill. To the left as you travel up the mountain are the Lillie and then the Frederika Mines. About a mile further the portal to the Silver Wing is to the right. In 1885 Crofutt felt the Lily and Golden Eagle were the two best mines in Burns Gulch. Both contained considerable quantities of brittle silver. Then a quarter of mile further are several old claims that were worked by the Great Eastern Mining Company in the 1940s and 1950s. The road dead-ends about two miles from its start into a basin holding the Golconda, Klondyke, Great Eastern, and other claims. A tough hiking trail goes over the divide and eventually to Cottonwood Creek on the Cinnamon Pass Road.

Back on the main Animas Forks Road, on the left side of the bridge over the Animas River, are the foundations of the Eclipse Smelter, which was built by James Cherry in the 1880s. It ran for several years and at one time had a three-story boardinghouse and office. It also had a steam-powered sawmill. The Eclipse Smelter was hit several times by snow slides coming out of Grouse Gulch. The smelter was operated by the Eclipse Mine in California Gulch above Animas Forks. All kinds of milling equipment were brought in and the huge operation employed a hundred men and used just as many pack animals. Surprisingly, at this altitude, the smelter did work, but financial problems set in, and the smelter operated at full capacity for only one month.

This end of the little valley that you are now entering once contained "La Plata City" (the area was in La Plata County at the time). N. A. Foss, H. C. Brown, and J. J. Epley began the San Juan Smelting Co. and platted La Plata City. Besides what was called "Brown's Smelter," they also had a small sawmill run off the same water wheel as the mill. There were no other structures in the "city." The whole operation was destroyed in an avalanche in March 1877, and the town site was abandoned. A little less than a mile after crossing the river there is a restroom where the road meets the Engineer and Cinnamon Pass roads.

The left fork goes to the settlement of Animas Forks, which lies at an elevation of 11,300 feet. Its original name was "Three Forks on the Animas" or sometimes just "Three Forks." A. W. Burrows and others built crude cabins here in 1873. Rhoda in 1874 found "several cabins with a number of mines about, who showed us specimens of ore from their various mines." By 1876 it had thirty cabins, a post office, a hotel, and two mills. The Dakota and San Juan Mining Co. mill and that of the San Juan Smelting Co. were built in 1876. The Mineral Mountain Mining Co. also built a concentrating mill in 1879. In 1876 the toll road from Lake City arrived at Animas Forks. The winter of 1876-77 was the first that anyone remained in town. In 1877 Animas Forks's population was estimated at two hundred in the summer and about sixty year round, and the town site was laid out that year. Lots in the town were free to anyone who would build a house or other structure. The town's earliest mill treated ore from the Red Cloud Mine in Mineral Point. In 1880 the *Ouray Times* described Animas Forks as "the business center

This is an early photo of Animas Forks, taken about 1880, before the big fire destroyed one side of town. Courtesy Denver Public Library, Western History Dept. (X6614)

The Gold Prince Mill at Animas Forks was a huge structure but it only operated for a few years before being dismantled and moved to Eureka. Courtesy Denver Public Library, Western History Dept. (X62192)

of Mineral Point and Poughkeepsie Gulch and always shows life, even in mid-winter." In 1882 Animas Forks had its own grade school. The wooden jail, which still stands, was also built about this time. The structure was made with two-by-fours laid sideways and is now located behind the Animas Forks restroom. Most of the buildings that still stand were built in the 1880s.

As you follow the road into Animas Forks, the Silverton Northern turntable was on the left just after passing over the Animas River (it is now only a large pile of dirt). It was installed in 1904, and most of the turntable's working parts were taken from the famous Corkscrew turntable on the Silverton Railroad. The house with the bay windows is often called the Walsh house, but it was actually built by mail carrier William Duncan (mail carrying was a good paying job!). On the left (above the town on the hill) are the remains of the Columbus Mill. The mine was located above the mill. It didn't close until about 1948 after being built about 1880.

Animas Forks occupied a strategic spot at the junction of the road to Lake City, the road to Ouray, California Gulch, and the road back to Silverton. In 1883 Animas Forks reached a population of 450 and even included its own newspaper, The *Animas Forks Pioneer*, which was founded June 17, 1882, and published until October 2, 1886 (ever notice how many places closed about the first snow of winter?). Ingersoll in 1885 called Animas Forks "another prosperous and populous mining area," but it was a different story in the winter. In 1884 the winter population was described as "twelve men, three women, and twenty dogs." Croffutt in 1885 stated, "It is in the midst of a wild and rugged country, where nothing but rich mines would ever induce a human being to live longer than necessary."

In the fall of 1891, a fire started in the Kalamazoo House hotel (originally built in 1881) and destroyed fourteen buildings in the town, because there was no water with which to fight the fire. Nevertheless, Animas Forks is still one of the best-preserved ghost towns in the San Juans. The town originally was laid out in a forest, but as time went by more and more trees were cut down, and avalanches occasionally hit the cabins. The snow can really pile up in Animas Forks. The *Animas Forks Pioneer* of March 1, 1884, reported many of the town's one-story buildings to be completely covered with snow, while snowdrifts in some places were seventy-five feet deep and it was still snowing! Seventy-five slides ran the next week on the road between Eureka and Animas Forks. By March 15 the newspaper reported the snow to be fifteen feet on the level. One blizzard that winter lasted twenty-three days and left twenty-five feet of snow on the ground. The only travel was on Norwegian snowshoes—today's skis. The Kalamazoo House could be entered only from the second story.

By 1900 the small settlement had almost become a ghost town, but when the Silverton Northern was extended to the town in 1904, new activity began at the Gold Prince Mine in Mastodon (Placer) Gulch. The town rebounded substantially, when the huge Gold Prince Mill (184 feet by 336 feet and one hundred stamps) was built in town in 1907. It was fed by a thirteen thousand-foot tram that sent ore down from the mine. It took four hundred carloads of steel to build the mill, including huge steel rafters that were incorporated into the roof. Reportedly the mill cost five hundred thousand dollars to build in 1904, but it ran for less than six years. The Gold Prince was the largest concentrating mill in Colorado at the time it was built. In 1917 the Gold King Mill was dismantled and sent to Eureka and the town again headed downhill. By 1923 Animas Forks was basically deserted, although several mines (including the Columbus, Early Bird, and Silver Coin) were worked nearby until well into the 1940s.

The roads beyond Animas Forks go three ways, all four-wheel drive only. One goes up California Gulch to Lake Como, another goes up toward then Engineer Pass Road, and the third goes over Cinnamon Pass. All of these roads are covered in detail in other chapters.

MAP OF INTERIOR OF THE SAN JUAN TRIANGLE

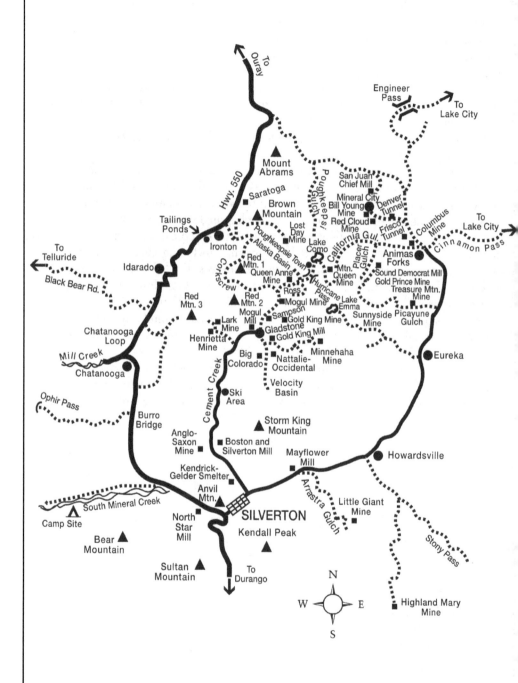

Chapter Nine

THE INTERIOR ROADS OF
THE SAN JUAN TRIANGLE

Corkscrew, Poughkeepsie, California Gulch

Within the core of the San Juan Triangle is a jumbled maze of short roads. The area is mostly above tree line (which in the San Juans is about 11,500 feet), but what this rugged land lacks in trees is made up by rich minerals. Many of the great mineral veins, and therefore many of the great mines of the San Juans, lie in this high alpine area that is accessible to vehicles for only three or four months in the summer.

Lower Cement Creek

The Lower Cement Creek Road is one of the easiest routes into these high mountains. It is passable by automobiles to the ghost town of Gladstone and traversable in fifteen or twenty minutes each way. The road passes several small mines and mills along the way. Most of the side roads go just a short distance from the main road. Tungsten was a metal found in abundance in many of the mines along Cement Creek; silver, gold, and base metals were also found. In 1878 a toll road was constructed from Silverton to Gladstone, and in 1879 it was extended up to the rich new discoveries at Poughkeepsie Gulch. H.A.W. Tabor, among others, purchased several mines at the head of Poughkeepsie Gulch, and a small six-ton mill was built at Gladstone to process the ore. The Poughkeepsie Gulch and Cement Creek roads then became a shortcut from Silverton to Ouray (but only on foot or on horseback) instead of the longer route via Mineral Point.

In 1899 the Silverton, Gladstone and Northerly Railroad was built along the seven and a half miles from Silverton to Gladstone. The little line provided a vital link between the D&RG Railroad in Silverton and the rich Gold King Mine in Gladstone. Otto Mears did not build the railroad, as many believe, but rather the Gold King Mine. Originally, the mine ran two trains a day to Gladstone. In 1910 Otto Mears, Jack Slattery, and James Pitcher leased both the mine and the railroad, and then in 1915 the Silverton Northern Railroad bought the Silverton, Gladstone and Northerly at auction. Just two years later the Mears partners gave up their lease on the mine. The railroad was officially abandoned in 1934, although it hadn't been used since 1924. Now the road generally follows the railroad grade, and the remains of the old road can usually be found nearby.

As one leaves Silverton the road curves around to the northwest heading up Cement Creek toward Gladstone on San Juan County Road 10. At about a quarter of a mile is the site of the Greene Smelter, Dempsey Reese's cabin, and the most prominent feature today because of its slag dump—the Kendrick-Gelder Smelter. Little remains of these early structures, but Dempsey Reese built his cabin in the area below the present-day slag dump in 1873 and filed on the town site of Silverton. Ironically, the cabin didn't end up located in the town, but Reese's

This photo, labeled "Silverton's First Smelter," was taken about 1890; it is probably the Kendrick-Gelder not the Greene. Courtesy Denver Public Library, Western History Dept. (X61438)

The Mogul Mill had been abandoned for some time when this photo was taken about 1940 but houses in town were still being used. Courtesy Denver Public Library, Western History Dept. (X8674)

house was used as a meeting hall and post office in 1874. The Greene Smelter was built in 1875 and was the earliest successful San Juan smelter. Burros brought in most of its construction materials and the mill's machinery piece by piece over Stony Pass. A small sawmill was also operated at the spot. The mill used water power from Cement Creek but was not very efficient; it closed in 1879 and its equipment was moved to Durango. The Kendrick-Gelder Smelter (also known as the San Juan Smelter) was built slightly upstream in 1890, and closed it in 1908.

The Gold Hub Mill was already showing its age when it had its picture taken in 1938. Courtesy Denver Public Library, Western History Dept. Walker Art Studio. (X62258)

About a mile further upstream the remains of a Silverton, Gladstone and Northerly trestle are easily visible in the canyon. The best vantage point (at mile 2.7 using the road markers) is at the turnout at the information sign structure, where you need to walk to the edge to get a good view. After another three-quarters of a mile the road crosses the creek, and a half-mile further are a retaining wall and the dump for the Mayday Mine (mile 3.8 on the road markers), which was discovered in the early 1900s but operated mainly in the 1930s and 1940s.

Shortly after the road crosses Cement Creek (mile 4.0 on the road markers), a mine tunnel and mill are visible across the creek. This is the Boston and Silverton Mill and the Yukon Tunnel. The hundred-ton a day mill was built around 1896 at the mouth of Illinois Gulch and was extensively remodeled in the 1980s. It was also called the Gold Hub Mine. The Yukon Tunnel accesses many different veins and was worked well into the 1980s. It processed a considerable amount of ore from the Uncle Sam Mine, located near the top of Storm Peak. About a half-mile further, the road to the left (County Road 61) dead-ends at about one mile into Ohio Gulch at the Queen City group of claims.

After another quarter of a mile or so on the main Cement Creek Road the Prodigal Son Mine is on the left. The Anglo-Saxon Mine (above the cribbing logs) is further up Porcupine Gulch. It was worked into the 1890s and early 1900s and then, most recently, was operated by the Gold King. The Anglo-Saxon produced tungsten. The old railroad grade is visible on the other side of the creek at this point. The big gulch on the right winds up Storm Peak Mountain, which has a microwave relay tower on top.

The Mammoth Mine (which can be seen to the left just after the road again crosses the creek) was started in 1901 in an attempt to intersect the rich Yankee Girl ores. It was a joint effort with the Red Mountain Mining Co. and was worked until well into the early 1900s. However, it was driven in only about two thousand feet, which was far short of the Yankee Girl. Today, there is also a sawmill operation at the Mammoth. The dead-end road continues past the Mammoth and leads into Georgia Gulch to the Kansas City Mine, which was located in the 1880s but was also developed later in the hope of intersecting Yankee Girl ore. Many attempts were made to reach the Yankee Girl, which gives you some idea how rich and well-thought-of the Yankee Girl ore was.

185

This view of Gladstone was probably taken from the Mogul Mill about 1940 and shows the Gold King in the background. Courtesy Denver Public Library, Western History Dept. (X8673)

There is an extreme ski area tram (Silverton Mountain) about a half-mile further. Then, on the left, after another half-mile (at mile marker 7), is County Road 35, which leads to the Lark and Henrietta Mines in Prospect Gulch. Here at the intersection of Prospect Gulch and Cement Creek was a "town" called Del Mine (also called Del Mino or Dela Mino). It had a post office from June 22, 1888, to May 19, 1884, and supposedly was platted and surveyed. The 1880 census gave it a population of twenty-five, and the 1884 *Colorado Business Directory* said Del Mino had a population of forty. However it never seemed to have been much more than a transfer point, where supplies were taken from wagons and loaded onto burros or mules.

After circling the face of the mountain on San Juan County Road 35 and entering Prospect Gulch, you will see the 5,300-foot Henrietta tram that ran down to the Silverton, Gladstone and Northerly Railroad. There are several claims in the basin. The Henrietta Mine is at the end of the first road to the left. The cabin and metal sheds to the right belong to the Lark Mine. The road to the left goes to the Crown Prince Mine, and the road to the right to the Joe and John Mine. All these mines (which are about a mile and a half up the gulch) were worked as late as the 1970s. Further up Prospect Gulch another road to the left leads to the Mammoth boardinghouse and then the upper workings of the Henrietta. The Henrietta and the Lark were located about 1905 and were owned by the San Juan Smelting and Refining Co. in the early 1900s. The ore was treated in Silverton. At the end of the road along the creek in the gulch is the Galena Queen Mine (don't take the road out of the gulch to the right because it is extremely rough and steep). The claim was located in 1890 and operated in the early 1900s. The Henrietta and the Lark were both worked well into the 1980s.

Back on the Cement Creek Road and right before entering the town site of Gladstone, the cement foundations of the Mogul Mill are to the left. The mill was constructed in 1906 and was connected by a ten thousand-foot tram with the Mogul Mine (discussed later) on up Cement Creek near the top of Hurricane Peak.

Gladstone and the Gold King Mine

A few people lived in the Gladstone area by 1877 but the town had its ups and downs. A major part of the problem was that the town was located at 10,250 feet. The post office as well as a general store and a hotel were established in 1878. The town was at first named Sampson but it was renamed for the famous British prime minister. Even though it only had a few cabins and a small school in 1878, it also had a small newspaper called *The Gladstone Kibosh*. In 1877 the machinery for a lixiviation (leaching) process was shipped to the site, but the mill was not actually working until 1878. It processed ore from the Red Rodgers Mine in Poughkeepsie Gulch and also ore from the Columbia Mine. The original mill was twenty stamps, then expanded to forty, and later to eighty. The Gold King Mine was discovered by Olaf Nelson in 1877. He was working at the Sampson when he discovered the rich Gold King vein nearby. The town had a small growth spurt about 1879 when rich strikes were also made in Poughkeepsie Gulch, and its population rose to about forty in 1880.

Crofutt in 1885 declared Gladstone to be "a mining hamlet on Cement Creek, way up in the mountains ... Population 100. General store, two quartz mills, a sawmill with several paying mines and many prospects in the vicinity compromise the town." By 1900 the population was up to three hundred because of the Gold King operation. However, Gladstone was so close to Silverton that the town didn't need a very big commercial district. It was basically a company town. The ore was eventually carried from the mine to the mill at Gladstone by a mile-long tram that was capable of carrying up to four hundred buckets of ore a day, each bucket carrying about 750 pounds. The ore was processed by a one hundred ton mill that was built in 1899 and expanded to two hundred tons in 1902. That year the Gold King

This view of the Gold King looks south into Velocity Basin. The speed skiing run was in the notch full of snow. Courtesy Denver Public Library, Western History Dept. C. R. Worthington Photographer. (Z2559)

9/380 GOLD KING TRAMWAY & GLADSTONE, COLO.

Both the previous photo and this one were taken for postcards by C. R. Worthington. Courtesy Denver Public Library, Western History Dept. (Z2561)

produced, milled, and shipped 72,455 tons of ore. Its mill had eighty stamps milling two hundred tons of ore with a recovery of seven hundred thousand dollars in gold. The mine expanded to over forty claims over the years.

By 1900 it was apparent that the Gold King Mine would be one of the largest in the San Juans. Two tramlines hauled ore down to the enlarged mill. When the new mill and the railroad were being built, it was reported that as many as two thousand people were sometimes in Gladstone, most of them living in several boardinghouses in the town. A huge four-story boarding house was constructed at the mine, high on the southwest side of Bonita Peak, and a scattering of small four-room houses and some smaller boardinghouses were built near the mill. The Silverton, Gladstone and Northerly Railroad was built by the mine at this time. In 1907 a fire destroyed many of the mine's surface buildings and smoke drifted into the mine initially killing two miners and then four more during an attempted rescue. The owners of the mine got in a fight over how to best develop it. As a result, Gold King Mine shut down in 1910, the railroad was abandoned, and the tracks were torn up in 1915. The mine reopened in 1918, and it was worked for many more years. When it closed in the 1940s the Gold King had produced about eight million dollars in ore.

What was later known as the American Tunnel was opened at Gladstone in the late 1950s by Standard Uranium Corporation, which was founded by Charlie Steen who had struck it rich during the uranium boom of the early 1950s. Overnight he had became worth millions when he discovered the famous Mi Vida Mine near Moab, Utah, and some of his money was reinvested in mines near Silverton. The purpose of the American Tunnel was to bore further into the mountain some six hundred feet below the previous lowest workings of the Sunnyside Mine, which had operated for decades on the other side of the mountain near Eureka. The Sunnyside had closed in the early 1940s when mining was directed away from gold toward the base metals needed by the United States for World War II. The Sunnyside Mine had gone so deep that it was becoming very expensive to hoist the ore and pump the seepage water to the surface. The new tunnel at Gladstone served to drain the workings by using gravity instead of expensive pumps. It was also hoped that the new discoveries might be found along the way as the American Tunnel was driven, which is exactly what happened.

The Mogul Mine was operating full blast when this photo was taken about 1910. Courtesy Denver Public Library, Western History Dept. (X62227)

Several major disasters occurred after the American Tunnel was driven. There were avalanches (the mine or mill were hit five times), floods, and the memorable event in 1978 when workers tunneled into the bottom of Lake Emma. The resulting flood did millions of dollars in damage, washed valuable ore and millions of tons of worthless mud for miles down the creek, and shut down the mine for two years. Although nine million dollars in property damage was done, there fortunately were no fatalities, because there were no workers in the mine on the Sunday when the flood occurred. Otherwise everyone would have been killed. A wonderful book full of miners' stories about Standard Metals (Sunnyside), as well as the Camp Bird, Idarado, Shenandoah Dives, and many other San Juan mines worked between the 1930s and 1980s is *Colorado Mining Stories* by Caroline Arlen.

Velocity Basin and Minnehaha Basin

Both of these trips lead a short way to dead-ends in beautiful basins above timberline. Behind the Sunnyside take the left fork towards the remains of the Lead Carbonate Mill, then in front of the old mill take the right fork south up County Road 52. It is an easy four-wheel drive trip along the South Fork of Cement Creek for about two miles to beautiful little Alpine Lake and Velocity Basin. The National Speed Skiing Championships have often been held here on this steep peak. About half way along County Road 52 are the Natalie-Occidental Mine to the northwest (left) and the Big Colorado Mine and Mill southeast across the creek. Both mines were worked in the late 1890s and early

This photo of the "town of Poughkeepsie" was taken September 9, 1901 by E. Howe of the U.S. Geological Survey. Courtesy Denver Public Library, Western History Dept. (X13089)

1900s. The Natalie-Occidental opened again during World War II. The tunnel was driven to reach the Black Hawk Vein at lower levels in Minnehaha Basin.

Also behind the Sunnyside and behind the remains of the Lead Carbonate Mill, you can go County Road 51 to Minnehaha Basin and the upper levels of the Gold King Mine. The forty-stamp Lead Carbonate Mill was constructed in 1947. The Lead Carbonate Mine is in Minnehaha Basin, and the ore was shipped down by truck. About six-tenths of a mile up from the mill is an avalanche-splitting device made out of old railroad rails and the tram to the upper levels of the Gold King. The road to the Number 7 Level of the Gold King is about a half mile further up. It leads off to the left, is usually gated, and often is not passable by four-wheel drive.

Another half mile up the road you enter Minnehaha Basin and come to a fork of three roads. The Black Hawk Mine is at the end of the right fork. It was located in the 1880s and worked well into the twentieth century. It was later worked from the lower tunnel at the Natalie-Occidental near Velocity Basin. The middle road goes to the Minnehaha Mine, which is in the upper part of the basin and was also located in the 1880s. It was operated by the Golden Monarch Mining Co. from the early 1900s into the 1930s. Later it was worked from the Roosevelt Tunnel, which was driven just below the basin. The left fork leads to the Lead Carbonate Mine, located in the 1880s by A.M. Jackson, but the main operations were in the 1930s and 1940s. Besides lead, the mine produced good quantities of silver and gold. It is also sometimes possible to travel another mile around into the steep basin that contains the upper workings of the

Gold King and the Sampson. The cabin and workings in the middle of the bend are part of the Gold King Extension Claim. There are several private homes in this area, so please respect private property.

Upper Cement-Hurricane Pass Road

As you leave Gladstone you can also travel on San Juan County Road 10 three miles to Hurricane Pass and Lake Como. You will also notice a road to the east of Cement Creek. This lower (eastern) road dead-ends at the Mogul Mine, which was connected back to Gladstone and its mill with a tramway. This is a rough four-wheel drive road, and most of the historic structures can actually be better viewed from County Road 10. Along the lower road are the remains of the Sampson Mill (the mine is much further up the mountain above the mill and was connected by a thirty-three hundred foot tram). The Sampson was eventually incorporated into the Gold King. Unfortunately, the original Sampson Mill was built in an avalanche path. In 1884 the entire mill and boardinghouse were destroyed when only four months old, but luckily only one man was killed. Tragedy struck again in 1900 at the same place. Four men coming down from the Gold King boardinghouse were going to the funeral of a friend killed in an avalanche. Just as they entered the Sampson workings they were hit by an avalanche, killing them when they were swept over a steep cliff. High above the Sampson in the gulch is the American Tunnel (Level 7) of the Gold King.

The Red and Bonita Mill was located slightly to the north (up the road) from the Sampson. It had a capacity of seventy-five tons a day. The mine was directly above the mill and had a three thousand-foot tunnel. It was discovered by the 1880s. It was owned by Joe Terry, son of John Terry of the Sunnyside Mine. The mill, which was built in 1899, had twenty stamps and a seventy-five ton capacity. Then further up the road several other smaller mines (the Adams and the Pride of the Bonita) are visible on the hillside and finally the large Mogul Mine is near the end of the road. The Mogul was located in the 1870s by F.M. Snowden. In the late 1870s it employed as many as twenty men, and a small camp was built at the site. Ore was carried back to its mill at Gladstone by its ten thousand-foot tram. The Mogul had extensive workings. Much of the upper part of San Juan County Road 10 is traveling through the mine. The Mogul tunnels eventually totaled twenty thousand feet, included fifteen claims, and extended all the way into Eureka Gulch where the Sunnyside Mine was located.

Two and a half miles from Gladstone the Corkscrew Road comes in from the west, then you encounter the loading bins of the Queen Anne Mine, which was located in the mid-1880s. These are the lower workings; the road passes through the upper workings for the next quarter of a mile or so. Its claims encompass the Adelphia, Columbia, and five others. Many of the claims in this area were owned by H.A.W. Tabor in the late 1870s and 1880s. Tabor paid $125,000 for the Adelphia. The Red and Bonita Mill treated the Queen Anne ore.

In another tenth of a mile a road leaves to the east into Ross Basin, which is situated between Bonita Peak to the south and Hurricane Peak to the north. This basin was prospected heavily in the 1870s but was worked the most during the early 1900s. It was also worked underground by the Sunnyside. Ingersoll reported in 1885 that Ross Basin was an "important" group of mines, which were worked by English capital and produced large amounts of bismuth. When the road forks in Ross Basin after about a half mile, the left side will take you to the Sunnyside Saddle or Whiskey Pass (the last part of which is very difficult and dangerous). You can travel only on foot into Sunnyside Basin. The right fork takes you to the shaft and the collapsed remains of a cabin on the Winchester Claim. Several beautiful little alpine lakes are also in the area.

Then back on the main road again, you travel toward Hurricane Pass on County Road 10. The road to the left leads down several very sharp switchbacks (be very careful if you try it at all) into Alaska Basin. The Alaska was owned by H.A.W. Tabor and is described further in the Poughkeepsie Gulch section of the Engineer Road Chapter. The ghost town of Poughkeepsie is located at the top of the divide.

After reaching the overlook of Lake Como, the main road skirts to the east and passes by the Bonanza Mine, whose various cuts and tunnels are on the hillside going down over Hurricane Pass. The Bonanza was discovered in the 1870s but was worked mainly in the 1880s and 1890s. The claims higher up Tuttle Mountain were the Silver Cloud, Starlight, Belcher, and Oberto. Down in the basin the road forks. One branch climbs up a steep talus slope to the southeast to California Pass (elevation 12,930 feet) and then on to Animas Forks down California Gulch. The other fork goes into Poughkeepsie Gulch and should be avoided because it is rough and dangerous.

Poughkeepsie City

At the top of Hurricane Pass is what little remains of the ghost town of Poughkeepsie. The town boomed about 1879 when H.A.W. Tabor came to the area and began operation of his Alaska Mine (to the north and lower than the pass). At that time the town grew to about 150 persons in the summer, but very few, if any, men were in the area in the winter. It had a store, restaurant, saloon, and many "comfortable" buildings. A post office was also located at the site from January 12, 1880, to September 15, 1881, and L.P. Kendall was postmaster. Kendall also started the town's newspaper— *The Poughkeepsie Telegraph*. Before 1879 Poughkeepsie's ores were usually packed out to Lake City. After that time they usually went to the mills in Gladstone. Crofutt, in his *Gripsack Guide to Colorado*, in 1885 reported, "The town was situated at the top of the divide and had a population of about 250 in the summer. It is the biggest little mining camp in the San Juan country." By 1890 the town was basically abandoned.

Poughkeepsie City seems to move around a little. It has been reported to be on the banks of Lake Como, here on the top of the steep slopes overlooking the lake, and at the other end of the gulch. Several photos show the town at the top of the pass, but there was also a settlement at Lake Como, so *Poughkeepsie City* could well refer to the whole area. The small store on the lake shore was usually called *Lake Como*. Roger Henn goes into quite some detail about the three Poughkeepsies in *Lies, Legends and Lore of the San Juans*. Its elevation was 12,050 feet and in a very windy spot, which made it most inhospitable, and hence the name *Hurricane Pass*. Major mines near the town included the Alaska, Saxton, Alpha, Red Rogers, Alabama, Poughkeepsie, and Pride of the Alps. Crofutt reported "the scenery about this area is most rugged and grand. Game is abundant, including grouse and quail, and an occasional grizzly bear." Grizzlies, by the way, were thought to be extinct in the San Juans until one was killed near Creede in the mid-1980s.

Corkscrew Gulch

The Corkscrew Gulch Road is a spectacular shortcut from the Uncompahgre River Valley over the Red Mountains to Cement Creek and down to the portal of the Standard Metals Mine at Gladstone or over Hurricane Pass to Animas Forks. The road was originally built by San Juan County in its attempt to access the rich Red Mountain strikes in 1882. The next year Otto Mears built a newer and easier route through Chattanooga, and the Corkscrew Route was forgotten. The old trail was made passable to vehicles in recent years to transport diamond-drilling rigs to various sites on the flanks of Red Mountain. The road is for four-wheel-drive vehicles only and should be avoided in very wet weather because of considerable areas of yellow clay that are extremely slick when wet.

At the north end of the tailings pond in Ironton Park on the Million Dollar Highway (U.S. 550), a small jeep road passes over Red Mountain Creek just below (north) the tailings. After passing by the tailings ponds, the first road to the left leads up onto Brown Mountain through Gray Copper Gulch. This is a great little side trip of about six miles, but it is a dead end because of a locked gate that keeps you from getting clear over to the Cement Creek drainage. Let's take the trip anyway.

As you travel up Brown Mountain past some good primitive camping spots, the first road to the left leads to the Saratoga and the Brooklyn, but it usually has a locked gate, because this is private property. These were both large mines that produced mainly low-grade ore but which were profitable because the Silverton Railroad terminated at their workings and their transportation costs were low. Then a quarter of a mile farther the road crosses Gray Copper Creek near a small modern settling pond, built as part of the local reclamation efforts. Half a mile farther, the road forks three ways. Take the right fork; the other two are private property and locked. About three-quarters of a mile farther are the remains of the Silver Mountain Mine. There are several well-preserved shacks and a fair amount of heavy mining equipment. A series of dumps reach up the hillside and belong to the same mine.

About a mile and three quarters further, the road enters a wonderful little basin with a pond and great views. The left

The Saratoga Mine and Mill were long abandoned when this photo was taken about 1910. Its brick smelter stack was used to create a ski lodge nearby. Courtesy Denver Public Library, Western History Department. (X62087)

fork into the basin is a dead end. The road to the right travels up to the edge of Gray Copper Gulch with some views of the Red Mountains and the dump and a building of the Lost Day Mine, but you might want to turn around at this point. The road continues to Gray Copper Falls (visible from here), but it is rough and without many spots to turn around easily. The road is gated just past the falls. There is also another gate blocking access from the other side.

Back on the main Corkscrew Road and traveling up Corkscrew, you pass above the reclamation work being done on the Idarado tailings pond. Tailings are the waste material from the mills, but because most mills don't recover all the precious metals, they often are high in minerals. The fine, dust-like material is usually mixed with water and pumped from the mill (in this case, several miles). Because the tailings are so barren, it is hard to get anything to grow on them. The water with which they are mixed often gives the tailings ponds the consistency of quicksand. Many millions of dollars have gone into trying to figure out how to make a tailings pond best grow vegetation.

As the road turns to the south around the tailings pond, take the right fork, because the left dead ends after just a few hundred feet. Then after about a mile of travel through the deep forest, the road comes to Corkscrew Gulch where the geology is yellow clay. The clay is very dangerous when wet, so you can use a road that forks to the left to skirt the area in wet weather. However, that road is only marginally better. It ties back in about four hundredths of a mile. If you look across the gulch just at the point where one first gets to the clay hills, the famous railroad-engineering site of the Corkscrew turntable can be spotted across the gulch, but you need to look carefully. Look for the railroad grade which dead-ends at the turntable; the turntable itself is only a few feet off the ground. The foundation and parts of the turntable still remain, although the building that covered it has long since disappeared. The Corkscrew Turntable provided the means of turning a locomotive and its cars around to get the train through a space that was too tight for a curve. It was the product of the inventive mind of civil engineer C. W. Gibbs, who worked for Otto Mears and also designed the Chattanooga Loop and the trestles of Ophir. The Silverton Railroad claimed it was the only railroad in the world with a turntable on its main track. The railroad used the turntable instead of backing the trains down the slope because the cars tended to derail or unhitch when the latter method was used. Gravity could be used to reverse a whole train-one car at a time.

Ascending up Corkscrew Gulch into the spruce forest, the road passes between Red Mountain #1 to the north and Red Mountain #2 to the south. The ruins of several old mine dumps are visible. The Midnight Mine is about a mile after the clay hills, right after you cross the creek the second time (in bad times you drive down the creek because it takes over the road). The midnight had good ore and was worked in the 1880s and 1890s. A boiler is still on top of the Midnight dump. About half a mile further you pass the remains of an old cabin on the right. It belonged to the nearby Earl Mine. There is also a modern restroom at this point. The road that loops around the basin leads to the Carbonate King, which was located in 1883.

Shortly thereafter, the road comes out into open talus slopes and begins a sharp ascent via a series of switchbacks to the 12,217-foot pass. The road to the right of the

The rider is on the old Otto Mears Cinnamon toll road. The Columbus Mine and Mill and the Frisco Tunnel are visible behind the town in California Gulch. Courtesy Denver Public Library, Western History Dept. (X6612)

summit goes nowhere-it just follows a narrow ridge out for some great views, but it can be very hard to turn around, so the best plan is to just walk out. Across the divide are several little lakes by the road. The Sailor Boy Mine is across the lake. It was mined in the early 1900s.

Descending toward Cement Creek you are afforded a great view of the basins below, pockmarked with adits and dumps from mines that waxed and waned over a century of fortune hunting. Most were connected with either the Mogul or the Gold King Mine that operated out of Gladstone for many years. There is a restroom about a mile down from the top. Shortly afterward there is a road to the left. It can be traveled for only about a mile past several beautiful little lakes and several mine dumps of the Extension Mine, ending at the Vernon Mine in Gray Copper Gulch. From this point it is private property so the road dead-ends, although it would tie into the Gray Copper Road from Brown Mountain if the gates weren't locked.

After a short distance further on the Corkscrew Road, the main road going down intersects one coming up from the former town of Gladstone (County Road 10).

California Gulch

The California Gulch road can be used in connection with the Corkscrew Route to gain access from Ouray to Lake City You can also use California Gulch with Cement Creek to get to Ouray from Silverton. All three routes are filled with the remains of old mines as well as the ghost towns of Poughkeepsie and Gladstone. I'll describe the approach to California Gulch from the Animas Forks side.

When you leave Animas Forks and travel up California Gulch on San Juan County Road 19, the Columbus Mill is just on the edge of town as you cross the West Fork of the Animas River. The Columbus was one of the largest mines in the vicinity of Animas Forks. It was discovered about 1880 and produced mainly low-grade zinc ore—however in very great quantities. It was worked hard until 1948 and has seen occasional mining since. Many of the cuts and tunnels on the hillside off to the west are a part of the Columbus Mine.

The next site encountered on the trip is the Bagley Mill. The huge mill building was prefabricated in California and erected at its present site "by the numbers" in 1913. The mill sits next to the Frisco Tunnel, which followed the original Bonanza Tunnel, later called the Mineral Point Tunnel. It was started in 1904 and eventually was driven some nine thousand feet into Houghton Mountain in an attempt to intersect many of the veins in the Mineral Point area at an elevation much lower than previously possible. It was supposed to be a large operation, which would incorporate almost all of the claims in the Mineral Point area. The tunnel was proposed to even extend over to the Poughkeepsie Gulch area, but it never actually made it that far. N. R. Bagley started work on the mill in 1904, as well as on the Frisco Tunnel, but the work went slowly. Early on, the tunnel accessed the ore of the Red Cloud Mine. In 1906 the tunnel was a thousand feet long, in April 1907 four thousand feet, and in April 1911 seventy-five hundred feet. In 1912 work on the mill was actually started. The arrival of the Silverton Northern Railroad at Animas Forks in 1904 had made the shipping of low-grade Mineral Point ore economically possible, yet the Frisco Tunnel shipped very little ore.

About a half a mile further along the route is the site of the Vermillion Mine and Mill (located above and below the road). The Vermillion was worked mainly around the turn of the century, and its mill was constructed in 1910. Its upper tunnels can also be seen higher up the mountain. The Burrows and the Ida are located a quarter mile further along the road. They followed the same vein as the Vermillion. All three mines were later worked out of the Bagley Tunnel. Fossett, in 1879, reported that the Burrows had "a galena vein of great size, but of rather low average grade-not so low, however, that the mine is not very profitable." It produced thirty-seven to sixty-five ounces of silver and sixty-five percent lead. Thirty to forty men worked a five-foot vein in 1878. The three mines were also worked at the turn of the century with some development under way as late as World War II.

There is often a lot of snow at the end of California Gulch, visible from this point, but the route is usually plowed and passable by the first part of June. The dumps you see as you go up the switchbacks belong to the Mountain Queen.

The Mountain Queen Mine's main shaft is four hundred feet deep and connects with a fifteen-hundred-foot tunnel, which was driven from lower in California Gulch. The mine was one of the first discovered in the San Juans and was shipping high-grade ore to the smelter in Lake City as early as 1877. The Mountain Queen sold for $125,000 in 1880 and was patented in1884. Handpicked ore was sent to Silverton and averaged one hundred ounces of silver per ton. The miners also worked the nearby Red Cross vein from a drift in the Mountain Queen. By the early 1890s Thomas Walsh leased and was working the Mountain Queen, Vermillion, and Ben Butler dumps and sending ore to his smelter in Silverton. The shaft of the Mountain Queen was sunk under the direction of Walsh. This is the only evidence found to offer any support to the story that Tom Walsh ever lived in Animas Forks. In a letter to his partner in Chicago, Walsh was a little discouraged and reported that, "the Mountain Queen pay streak does not hold its value going down. For 600 feet on the surface it averages 100 ounces. Four feet down it falls to 30." The mine was also reported by Frank Fossett to have "a galena vein of great size, but of rather low average grade."

In 1901 about twenty men were working a rich strike that was six feet wide, sixty-five percent lead, thirty ounces of silver, and contained a small amount of gold. In 1904 Rasmus Hansen worked the mine on a lease, then bought it for twenty-five thousand dollars and sold it just a little later for one hundred thousand dollars. It was worked off and on from that time.

California Pass is 12,930 feet in elevation, then the road drops into Poughkeepsie Gulch at Lake Como. The road through Poughkeepsie Gulch is passable by four-wheel drive vehicles for only a short distance and then becomes extremely rough, so go up the south side of the basin over Hurricane Pass (12,407 feet) to the upper Cement Creek drainage.

Placer Gulch and Gold Prince Mine

Placer Gulch was also known as Mastodon Gulch after the Mastodon vein and mine. This fascinating road can be accessed from either the Silverton-Animas Forks Road or from California Gulch on San Juan County Road 9. The road circles around Treasure Mountain through Picayune Gulch and Placer Gulch. It is usually filled with brilliant wildflowers. I'll describe the trip coming from the north and going south from California Gulch.

As you enter the gulch, the remains of the Gold Prince Tram are very evident on the east side of the valley. The tram ran nine thousand feet from the upper end of Placer Gulch to the angle station (the base of which is the large square structure high on the hillside to the left) and then another four thousand feet down to Animas Forks. The total length of the tram was thirteen thousand feet. It was built in 1905 at a cost of seventy-five thousand dollars for the purpose of bringing Gold Prince ore to Animas Forks after the arrival of the Silverton Northerly Railroad. The huge Gold Prince Mill was in Animas Fork,s and the tram brought the ore down at a speed of about four miles an hour (fifty tons per hour).

The Evening Star is the first major mine with some buildings still standing. It is on the right about a half-mile into Placer Gulch. Most of the mine's work was done about the turn of the century. Another quarter of a mile further and across the valley is the Index Mine. After about another three-quarters of a mile, the buildings in the valley by the creek are first the Mastodon Mill and then the Sound Democrat Mill. The Mastodon Mill was constructed in 1885 and treated ore from the Mastodon Mine at the very head of the gulch and on the right. Not much of the Mastodon Mill remains.

The Sound Democrat Mill is above it and was also connected by an eleven hundred foot tram to its mine, which is directly south on the hillside. The Sound Democrat Mill was built in 1906. It processed ore from its mine, which was first extensively developed in 1899. The mine had very rich ore—some samples running over a thousand dollars

This is an early view (before the tram) from the Gold Prince Mine looking north down Placer Gulch toward Animas Forks. Courtesy Denver Public Library, Western History Dept. (X62190)

a ton. The mill was expanded in 1909 to also treat ore from the Silver Queen Mine. The Silver Queen was discovered in the 1880s and was developed on several levels. The mine sold in 1901 for twenty-five thousand dollars.

The Sound Democrat Mill has been restored and is well worth stopping for a look. It is one of the most complete nineteenth-century mills in the San Juans-a classic example of a small concentrating mill, which was used to reduce the ore by taking out as much of the worthless rock as possible. This process greatly reduced shipping costs from the remote area. The mill was built in 1905-06 and had four stamps, then three more stamps were later added. The mill was originally powered by steam and later by electricity. After the ore was crushed, it was spread over the Wilfley tables near the bottom of the mill. The heavier metals sank and were caught by the ridges or ripples. Periodically, the concentrates were collected from the ridges, dried, sacked, and then sent by pack mules or wagons to Animas Forks for shipment by the railroad to the smelters. Between 1909 and 1914 the mill produced only about twenty thousand dollars in precious metals from the Sound Democrat and Queen Anne Mines because it was worked only sporadically, but at today's prices the ore would be worth almost a half million dollars. It closed in 1914 and its owner, Joe Terry, helped develop the nearby (as the crow flies) Sunnyside Mine, which came to be one of the greatest gold producers in Colorado. The three mines that you see about halfway up on the hillside directly ahead and tied together by an impassible road are, right to left, the Hidden Treasure, Silver Queen, and Sound Democrat Mines.

Sometimes there are very complete brochures at the mill that describe the processes, but in case there are none the processing of the ore was as follows. The ore came in from the tram at the top of the mill. The water tank stored the water that was used during the milling. The water came down a wooden flume from a concrete dam that can still be seen upstream in the creek. Then the ore was dropped down the ore chute to the jaw crusher, which reduced it to about the size of ping-pong balls. These small pieces went to the stamps, which reduced it to powder and then to the amalgamations plates and finally to concentrating tables. The dried concentrates were sacked and shipped by wagon to the railroad at Animas Forks.

As you reach the end of the south end of Placer Gulch, you are entering the various workings of the Gold Prince Mine, which was also called the Sunnyside Extension (since it is an extension of the rich Sunnyside vein in Eureka Gulch). On the right, a short road leads to the Gold Prince boardinghouse's concrete foundation. It was a large structure, built in the early 1900s, and could hold 150 men. At this point there is a road that leads a short distance to the west (right). It goes directly around the Hansen Mill (also called the Sunnyside Extension Mill). The twenty-stamp mill was built in 1899 and could process forty-seven tons a day. Rasmus Hansen discovered the Sunnyside Extension in 1882 and ran it until 1903. The Mastodon vein was huge (75 to 150 feet wide) and rich. It is the same vein as the Sunnyside, which is an hour away by jeep, but just over

S. G. White took this view of the Sunnyside Extension Mill about 1895. With all the women and children it may have been at the opening of the mill. Courtesy Denver Public Library, Western History Dept. (62199)

the hill as the crow flies. The Neptune Mine was also on the Gold Prince or Mastedon vein. A small tram was built in 1893 to bring the ore from the upper levels down to its mill, which was called the Hansen Mill after its owner Rasmus Hansen. There are also the remains of a smaller, older mill in front of the Hansen Mill. The road (path) continues on to the Neptune Claim.

The main road continues uphill past more workings of the Gold Prince Mine. The mine was discovered in 1874 by Morris Rich and was originally known as the Mountain Pride, then the Sunnyside Extension, and then in 1903 the Gold Prince when it was sold by Rasmus Hansen to Cyrus Davis and Henry Soule who owned the Gold King in Gladstone and built the mill in Animas Forks. Later the Sunnyside operated the mine from underground, bringing the ore out at Eureka. It was a very rich mine, shipping some ore worth as much as three thousand dollars per ton.

The road continues up. Forget the road to the right—it is difficult and only goes a short distance to additional Gold Prince workings. The main road then passes the tram station for the Gold Prince Mine. For the next mile the road is going through the upper workings of the Mastodon, the Gold Prince, and the Sunnyside Extension coming in from Eureka Gulch. Near the top of the pass there are some great views—not just of Placer Gulch but also of the entire San Juans. The turnout to the south (right) lets you look into Parsons Gulch back toward Eureka Gulch and the Sunnyside Mine. Then you are skirting Treasure Mountain (to the northwest) and start down into Picayune Gulch (*picayune* means *worth a small amount* or *nearly worthless*). The mines in this gulch were not that bad, but there were no extremely rich claims.

At this point dumps down to the right are part of a modern-day operation, and dumps on the left go with the Golden Fleece Mine and the Scotia Mine (the lower mine). These mines were located in the 1890s. They were bought and became part of the Gold Prince Mine when Rasmus Hansen owned it. Both claims were worked into the 1940s, and they produced extremely rich gold ore—in fact some of the richest free gold in Colorado.

After a couple of switchbacks (and about three miles from starting this side trip) the remains of the San Juan Queen are to the south. It was located in 1890 high in Picayune Gulch and yielded small amounts of very high-grade gold and silver. Rasmus Hansen also worked these claims in the 1890s. About the turn of the century a small boardinghouse and other buildings that went with the mine were built.

Shortly after this point there is a loop road that comes around from the other side of Picayune Gulch. It is a rough road that passes through very little of interest. On the main road a little way further is the Treasure Mountain Mill and Mine on the east (downhill) side of the road and almost four miles since the start of the side trip. It was a large operation started in the early 1900s and composed of some thirty claims. Because it was worked fairly recently a little more remains of its buildings than at the older mines in the area. In 1937 this operation started exploring the Golden Fleece and the Scotia at lower levels by driving the Santiago Tunnel. Unfortunately, like many places in the San Juans, the ore lost its value as it went deeper.

The next road to the right is the other end of the loop road described above. The main road descends quickly and steeply to County Road 10—the Animas Forks/Silverton road. The whole trip is just short of five miles and takes about an hour.

The Mineral City (Mineral Point) Road

As one leaves Animas Forks traveling north, the first road to the right (after about a quarter of a mile) is the Cinnamon Pass Road, described in another chapter. A short distance further there is a mine across the canyon with cables still connecting its tram to the road side of the canyon. There are also several cuts and small dumps on the hill above the road. Mine cables like this have caused several airplane crashes in the San Juans; a helicopter crashed at this site, but there were no serious injuries.

About a mile further you come to a road to the left (southwest) that can be traveled for two miles and dead-ends into the town of Mineral Point (also called Mineral City). This early mining camp was first settled in 1873 (prior to the founding of Animas Forks) and

The town of Mineral Point hides in the trees soon after it was founded in 1873.
Courtesy Denver Public Library, Western History Dept. Barnhouse and Wheeler Stereo Card. (X12327)

continued to be active about 1890. A half mile along this road we encounter the Denver Tunnel, which was built to access the many low-grade veins on the hillside behind the tunnel. The site still contains a cabin, loading chute, and a good-sized dump, as well as the foundation of a small concentrating mill. About a hundred feet further down the road is a large cut into the side of the mountain, which exposes one of the many low-grade mineral veins in the area. The remains of the electric and telephone lines running to Mineral Point are also visible. It is hard to believe that such amenities were available up here in the 1890s, but they were!

Then about three-tenths of a mile further the remains of the Red Cloud Mine are visible in a small basin across the creek. A short dead-end road leads to a nice head frame, which was used for the deep shaft, and the winze that wrapped cable around a spool and raised and lowered the mine's bucket. Someone, as part of a reclamation effort, has planted trees in the area. At first glance it looks like a graveyard because of the shingles stuck in the ground to protect the trees. However, the whole effort seems a little futile because the area is above timberline. Up above there is another shaft that still has its boiler and winze. There are some great views in this area—you can see forever—but please stay on the road and don't destroy the tundra.

About three quarters of a mile further on the main road is the actual site of Mineral Point and the Bill Young Mine. It was first known as Burrow Camp, then Mineral City, then Mineral Point. Enough traces of the streets, foundations, and buildings remain to locate the settlement upon a close examination, but it was spread all the way from the mine at the end of the road, out into the high ground in the swampy areas (it was not swampy nine months of the year). Charles McIntyre and Captain Albert W. "Abe" Burrows built several cabins at Mineral Point in 1873. They also located the Big Giant, Bill Young, Red Cloud, Boston, Burrows, and Bonanza nearby. Ed Tonkyn was later the postmaster, mayor, hotel operator, and deacon of the church at Mineral Point for many years. Mineral Point is one of the highest ghost towns in Colorado at 11,700 feet. The town is named for a sixty-foot knob of quartz with silver veins that was just to the east of town. The vein (which the knob is a part of) can be seen running across the entire valley. There are many low grade veins visible in the area. The cabins were widely scattered around the flat, marshy area, but six or eight were close enough that they might be called "the town." Most of the cabins ringed the marshy area.

By 1875 a real "town" was under way at Mineral Point with twenty cabins and fifty to seventy-five residents. The town grew again in 1876 with several new stores, a justice of the peace, restaurants, and a mining brokerage. The first representative to the Colorado Territorial Legislature from the Silverton-Ouray area was a man named McIntire from Mineral Point. Although Mineral Point was already well established and was the jumping off point for the prospectors who came to the site of Ouray in 1875, it didn't get a post office until a day later than Ouray. Things were going well enough at Mineral Point that the *Ouray Times* reported that thirty-two miners spent the winter of 1876-77 at Mineral Point working their claims. The concentrating works of the Decatur and San Juan Company were located here. Although Mineral Point was described early as "a motley assortments of tents and a few poorly built cabins" in 1877, it was "thirty well-built cabins, all with the latch strung on the outside." By the end of the summer of 1877 there were forty buildings and a sawmill (which was worked until 1882). Seven women resided at Mineral Point that summer. The Forest House was the best-known saloon. Mineral Point's post office had one brief closure on October 14, 1878, but was reopened after the harsh winter on April 23, 1879 and continued operation until January 28, 1897. Before the construction of the Million Dollar Highway the town was an important summer stage stop, since it was located about halfway on the Silverton to Ouray road. Note the corduroy road (made of logs laid side by side) through the swampy area so that stages and wagons could travel this route. The peak population in the summer might have reached several hundred hopeful prospectors, but the winters were so fierce that only a few diehards remained in the town. The U.S. census officially listed 163 inhabitants in Mineral Point in 1880.

The Mineral Point Hotel opened in 1882. It was twenty-four by seventy feet and made of sawn lumber. Several people spent the winter at the hotel during the 1880s, and the hotel averaged about fifty boarders or travelers in the summer. Later it became a rooming house. When the town of Poughkeepsie was closing down in the mid-1880s, Mineral Point took up the slack. Crofutt states in 1885 that Mineral Point was still a promising mining camp. "A large amount of ore is taken out here and hauled on wagons to Lake City for reduction, which ran $100 to $600 per ton...The town consists of one store, saw mill, several restaurants and saloons, and the requisite number of cabins for a population of 200."

The town of Mineral Point was nicknamed "the Apex of the Continent" (even though it was not located on the Continental Divide) because the Animas River, Uncompahgre River, Lake Fork of the Gunnison River and Henson Creek all have their headwaters within a mile of the townsite. These rivers eventually drain a major portion of the San Juans, hence the name. A local justice of the peace, when threatened by an appeal of his decision to a higher court, is reported to have replied, "There is no higher court." By the turn of the century, the area was completely deserted except for a few summertime prospectors. The ore was too low grade, the spot too remote, and production costs too high to keep the mines open.

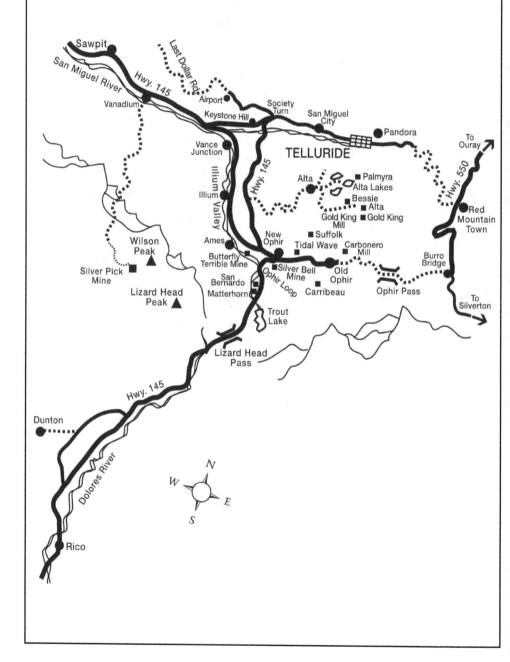

MAP OF TELLURIDE, OPHIR, AND RICO ROADS

Sawpit

San Miguel River

Hwy. 145

Last Dollar Rd.

Vanadium

Airport

Keystone Hill

Society Turn

San Miguel City

Pandora

To Ouray

Vance Junction

TELLURIDE

Illium Valley

Hwy. 145

Hwy. 550

Illium

Alta

Palmyra

Alta Lakes

Bessie

Alta

Gold King Mill

Gold King

Red Mountain Town

Ames

New Ophir

Tidal Wave

Suffolk

Carbonero Mill

Butterfly Terrible Mine

Silver Bell Mine

Old Ophir

Burro Bridge

Wilson Peak

Silver Pick Mine

San Bernardo

Matterhorn

Ophir Loop

Carribeau

Ophir Pass

To Silverton

Lizard Head Peak

Trout Lake

Lizard Head Pass

Hwy. 145

Dunton

Dolores River

N

W

E

S

Rico

⌒ *Chapter Ten* ⌒
TELLURIDE TO RICO

There are several exciting four-wheel drive trips out of Telluride, and several short side trips that can also be made, in addition to the all-weather road (Highway 145) to Rico. Imogene Pass and Tomboy Basin are discussed in another chapter. The Black Bear Road begins to the east of Telluride, but is one way coming from the Million Dollar Highway on Red Mountain Pass to Pandora, so we will start by heading east out of Telluride for only the short distance to Pandora.

The little town of Pandora contained the Pandora Mill, two Smuggler-Union mills, and the Tomboy Mill. The ore came pouring into Pandora from the nearby mountains by railroad tram, wagons, mules, burro trains, and aerial trams. The Pandora Mine was discovered in 1875, and the nearby town was platted the next year under the name *Newport*, for Newport, Kentucky-home of one of the founders. It was also at times called Folsome—the name of the company that ran several of the nearby mines. The name was changed to Pandora in 1881. The trail to the Smuggler and Tomboy originally started at Pandora, and the town was after 1892 the terminus of the Telluride Branch of the Rio Grande Southern Railroad. The Pandora Mill was built in 1881 with forty stamps. The Smuggler-Union also had what was called the Red Mill, built in 1899. It had sixty stamps and was connected to the Bullion Tunnel by a tram, which replaced the Sheridan Incline. In later years the Red Mill also serviced the Black Bear Mine. The Smuggler-Union Mill was eventually operated by the Idarado Mine almost a hundred years later. The Liberty Bell, Tomboy, Black Bear, and Mayflower Mines also had ore-loading facilities at Pandora. About the turn of the century Pandora had a Methodist church, a school, and about thirty homes, but as the ore slowed down the population of Pandora fell to only thirty-five in the 1930s. Nearby Telluride felt blessed that the mills, with their accompanying noise and soot, were a few miles down the valley.

East of Pandora there are two waterfalls—Ingram to the north and Bridal Veil to the south. Bridal Veil at 365 feet is the tallest falls in Colorado. The road cannot be traveled to the east beyond Bridal Veil. The legend of the falls is that honeymooners once visited the power station. The new husband slipped but grabbed a rock. As his bride was reaching for him, he fell to his death on the rocks below. The hydroelectric plant (it still operates) and house at the top of the falls were built in 1907 to provide electricity for Pandora.

A mile west out of Telluride is the site of San Miguel City. Newer buildings have been built there, but nothing remains of the original town. San Miguel City was located on Mill Creek and was founded by F. P. Brown, J. H. Mitchell, and Thomas Lowthian in August of 1876. It was surveyed in 1877 and contained about a dozen cabins. By 1879 it boasted about a hundred inhabitants. Crofutt wrote in 1881 that "it contains several good stores, a hotel, two stamp mills, and one concentrating works," which were for the nearby placer mining operations. It had a school with several dozen children attending. Its population was about two hundred and Crofutt verified that "this country is surely

This view of Pandora shows the Smuggler-Union Mill, the Red Mill, the RGS loading facilities and the old trail to Marshall and Savage Basins. Author's Collection.

destined to become the most important placer mining portion of the state, if not the world... as almost, if not quite, every foot of ground is gold-bearing." Pandora had a newspaper in 1881-82 and again in 1886-88. By 1895 Hall mentioned that a great number of the population of San Miguel City had been "absorbed" by Telluride.

The turnoff to Rico (Colorado 145) is named Society Turn because the upper class of Telluride used to come here for picnics, to socialize, and to be seen by others. But for now let's continue west into the Ilium Valley. By continuing down Keystone Hill (Keystone was a small placer mining community at the very west end of the valley), you come to the turnoff into the Ilium Valley. It is not only a very beautiful place, but also a valley full of history. The Keystone Placer was on the hill just above where the San Miguel and the Lake Fork join. It was first operated in 1881, and soon large flumes were built and hydraulic mining was started. It operated off and on until 1896, but was never really a paying proposition.

Just a little bit up the valley, a spot named Vance Junction once existed across the river. It was named after Jim Vance, an early prospector who had a cabin at the spot. Its population was twenty-eight in 1900. It was the junction of the spur track that led off the main line of the Rio Grande Southern Railroad to Telluride. The train left the main track at Vance Junction but then kept going upstream where it made a big loop near Ilium and came back downriver to climb Keystone Hill. A couple of coal tipples still stand on the hillside across the creek at Vance Junction because there were coal mines in the area, and the locomotives restocked at the spot. There used to be a two-story agent's home, a railcar which was initially used as a depot, and a post office, which was moved back and forth between this spot and Ilium.

About a half-mile further is Ilium, which is today an Episcopal Church camp but had its beginnings as the site of the second hydroelectric plant in the San Juans and a place where much of the early day training on electricity was conducted. The powerhouse buildings still remain at Ilium, but the generator, Pelton wheel, and other power-making machinery have been dismantled and moved away. Later an eight-car spur, section house, the superintendent's house, and a wye for turning the engines coming from Rico and going to Telluride were also located at Ilium. Most of the railroad grade can still be seen on the hillside above Ilium.

The Butterfly-Terrible Mine was located four or five miles further near where the Butterfly Trestle crossed the Lake Fork of the San Miguel River and just to the west of Ames. The Terrible Mine was discovered in 1883 and was located uphill from the Butterfly. Eventually the two mines came under the same ownership and employed thirty to fifty men during their peak production. Its thirty-ton mill was on a five hundred-foot spur from the main RGS track, and a tram brought ore from the mine, which was about halfway between the upper and lower tracks on the Ophir Loop. The mine was not a major factor until 1898, when it began producing significant amounts of both silver and gold. The original mill burned in 1927, and another was built. In 1940 that mill also burned, and the mine closed in 1953. Most of the area has now been absorbed into a new subdivision or destroyed by snowslides. The Butterfly-Terrible produced about a million and a half dollars in ore.

Next is Ames, originally a little mining and milling town on Howard's Fork of the San Miguel River. A post office was established in Ames on December 20, 1880, which continued to operate until June 1922 with only a few short interruptions. Ame's Boston and Ophir Smelter was the first smelter in the San Miguel County (built in 1882), but it failed for lack of good coal and the resistance of the local ores to its processes. It ran a total of only seventy-five tons of ore in 1882. Ames also had a newspaper (the *Ames Argus*) during 1883-84. The 1884 population was about a hundred, but by that time its days as a mining camp were numbered.

Ames's chief claim to fame came on June 21, 1891, when it became the first spot to generate and transmit alternating current over long distances for commercial purposes. L. L. Nunn, whose brother was an engineer at the Gold King Mine, built a crude shack and installed a six foot Pelton wheel with an attached generator. It dropped power costs at the Gold King Mine from $2,500 per month to $500. Water flumes ran from Trout Lake to both Ames and Ilium to power the generators' turbines. Until that time, all generators, motors, lights, or appliances used Thomas Edison's direct current, were designed for commercial purposes only, and could be economically transmitted over only short distances. A school, called the Telluride Institute, was established for engineers who were interested in electricity. Students came from all over the United States. The locals called them "pinheads." Ames was a stop on the RGS line because there was a water tower at the spot.

A motor was installed at the Gold King Mine at Alta, and Telluride was the spot of the first use of alternating current in 1891. The Ames electric plant soon expanded its generating capacity with the addition of the Ilium plant and transmitted power to Telluride and many of the other towns and mines in the area-some as far away as the Camp Bird and Revenue Mines located near Ouray. Telluride was one of the first cities in the world to have electric streetlights. The little electric power plant at Ames still generates electricity.

One of the great San Juan legends occurred along this stretch of river. On September 6, 1909, a flood occurred when the dam broke at what was known as "Middle Basin" halfway between Trout Lake and Lake Hope, and located at twelve thousand feet. Supposedly pinheads at Ilium and Ames were warned of the flash flood by telephone and they calculated on their slide rules how high the water would rise at their locations. They retreated only that distance, where they saw the flood waters come to just inches of their calculations. The flood washed out roads and the RGS tracks, and it was many months before normal travel was restored.

The Rio Grande Southern Railroad caused economic disaster for the town of Ames. Its tracks were about four hundred feet up the mountainside from Ames, so many Ames businesses and residences moved a mile further upstream to New Ophir, leaving only the power-generating plant and its employees at the site.

In 1891, when the Rio Grande Southern Railroad was being constructed, a bit of fantastic engineering was performed near Ames. What came to be known as the Ophir Loop crossed the canyon and headed back up again to gain enough elevation and move on to Trout Lake and Lizard Head Pass. "The Loop" required three large wooden trestles on its lower part, then four more above Ophir on what was called "The High Line," as well as an enormous amount of blasting of the hard rock in the area. It became an engineering marvel and a railfan's delight (one entire five hundred page book in the Sundance Books' RGS series is devoted to just the Ophir Loop and the nearby areas.) Like in Ames, most businesses in Ophir moved two miles down the creek to a location that became known as New Ophir. The original town of Ophir then came to be known as "Old Ophir."

At the point where the Illium Road ties in with Colorado 145, at New Ophir, you can turn left and return to Telluride, turn right and go to Trout Lake and Rico, or go straight two miles up Howard's Fork to Old Ophir and on to Ophir Pass. Ophir had a depot and served as a spur to the Silver Bell Mine, which was located directly above and south of the town. It also had a facility for loading ore that came by tram from the Gold King Mill at Alta. The Silver Bell Mill was also established at New Ophir, as well as

D & RG boxcars have been left at the Butterfly-Terrible Mill in 1938. Courtesy Denver Public Library, Western History Dept. (X62291)

Before the arrival of the R.G.S. Ames had a sizeable little community. Courtesy Denver Public Library, Western History Dept. (X62291)

two or three dozen houses and several businesses, including a newspaper, general store, blacksmith shop, livery stable, several saloons, a hotel, a post office, and the RGS depot. The post office at Ophir was originally called Ames, although that settlement was quite a ways from the post office. After the post office closed at Old Ophir in 1921, New Ophir got to use the name Ophir. However, the RGS had always called New Ophir just "Ophir." Only the old Oilton Club (a café and bar) building and some of the Silver Bell buildings remain today. The depot was torn down and its site lies under the highway loop.

The whole Ophir area was often the scene of heavy snows and avalanches (the elevation was 9,300 feet and obviously the mountains nearby are steep!). After an avalanche hit the first depot, a new depot was built in a less avalanche-prone area about a thousand feet downhill in town. Ophir's depot became one of the most photographed in the entire world. Besides the trams coming in from the Silver Bell and the Alta, many packers brought ore down by mules to be shipped out of Ophir.

The Silver Bell Mine was located in 1879, about thirteen hundred feet above where the town of New Ophir would be founded. The Silver Belle produced mainly silver (by 1882 it was averaging 190 ounces per ton) and some gold and lead. Early on it shipped its ore over Ophir Pass to Silverton to be milled. In 1901 it built its twenty stamp mill at New Ophir, which was quickly expanded to fifty stamps. It shut down because of disagreements among its owners in 1906, then produced only a small amount of ore until 1922 when it started producing well. The Silver Belle was worked with the Butterfly-Terrible through underground tunnels in 1937-40. In 1951 the Silver Bell also operated the Carbonero out of Old Ophir. It was worked until 1976, so much of the mill is still visible at Ophir.

The two and a half mile tram from Alta came down right beside the depot so the ore could be dumped directly into the gondola cars of the RGS after 1910. High-grade ore was loaded into freight cars to protect it from theft. An ore bucket arrived every two and a half minutes from Ames when the tram was running.

If you follow paved Highway 145, about a half mile back toward Telluride, you come to a dirt road that you can follow up to the old mining camp of Alta. The road is easily traveled by a jeep. There are some primitive camping spots along the way, as well as great views of the Ilium Valley. A little less than four miles up the road you will encounter the Gold King Mine and the town of Alta. The well-preserved remains are the site of the old mining camp—never really a town but nevertheless a sizeable community. The post office department called the Alta office Sultana since there was already an Alta in California. Alta means high in Spanish and the town was located at eleven thousand feet-very close to timberline. The Alta Mine was located on nearby Silver Mountain. Fossett reported that the mine was producing up to 290 ounces of silver per ton. A boardinghouse, school, cabins, and a forty-stamp mill were all in the area.

The Gold King was the biggest early-producing mine in San Miguel County. In 1881 the Gold King also built the first stamp mill in the county. It was ten stamps, brought from Alamosa with forty-five yoke of oxen pulling the wagons. Ten more stamps were added in 1882. The first ten sacks of ore from the Gold King Mine brought five thousand dollars. It was working a four-foot pay streak, which averaged about fifty dollars in gold per ton before milling. In 1882 the Gold King worked fourteen men. In 1884 fifty thousand dollars in gold bullion was shipped from the Gold King's twenty stamp mill, and by 1890 the mine was producing about two hundred thousand dollars a year. As mentioned earlier, high production costs at the Gold King Mine were responsible for the first alternating power plant being built at Ames.

The Ophir Loop of the Rio Grande Southern Railway held the little town of "New" Ophir about 1925. Author's Collection.

The big building that still stands at Alta was built in 1939 as the boardinghouse for the Alta Mine. The house to the left was the assay office and company store. John Wagner operated a mill near this site from 1917 to 1948. The mines around Alta eventually produced over five million dollars in gold, silver, lead, and copper. There were two main trams coming into Alta—one from the Palmyra crushing plant near Alta Lakes and another from the Alta Mine in Gold King Basin. The Gold King Mine had its own mill on site, and then there was another tram to ship the ore down to Ophir Loop. Many of the buildings that still stand were housing for the mine's married employees, and Alta had a school for their children. The Belmont-Wagner Mill stood in the big basin that the road loops around. Most of the mining in the 1920s and 1930s was done through the Blackhawk Tunnel, which was right next to the mill.

The road behind the gate is public, but please respect private property. The remains of many of the old cabins and other buildings can still be made out, although some are badly deteriorated. The road forks after a short distance. The left fork is the Boomerang Road that ties in with the Telluride Ski Area, but it is usually locked. The right fork is very rough and leads to the Alta Lakes. It is much easier to get to Alta Lakes on the main road.

After about a quarter of a mile the road forks with the left fork leading to the lovely lakes and a campground with restroom facilities. The first explorers found the fishing good at Alta Lakes, and it is still good fishing in all three Alta Lakes, as well as being a favorite short side trip for jeepers and photographers. The remains of buildings that exist above the upper Alta Lake are what is left of the Palmyra Mill, which was connected by track to its mine further uphill. The Palmyra was operated by the Alta after 1913. The St. Lewis was another mine on the lakes.

The right fork leads a half-mile into Gold King Basin. The Gold King was located in 1878 by Jack Mann, and it produced a gold ore that was easily milled. The Gold King had the first mill (1881) in the county. After 1882 a half-mile aerial tram brought the ore from the mine high up in the southeastern part of the basin to the mill. Later a lower tunnel accessed the mine and a large trestle connected the mill and the mine. In 1890 the Gold King yielded two hundred thousand dollars in four months. The mine's ore had little in the way of base metals—just gold. By 1891 the Gold King was one of the most important mines in San Miguel County; however, it was worked very little after 1900. Altogether the mine produced over $3 million in ore. As explained earlier, the Gold King's main claim to fame was its use of alternating current, and it prospered greatly under the management and ownership of Lucius Nunn.

The Alta and Bessie Mines were on the northern side of Gold King Basin. The Alta was discovered July 25, 1877, by Pat Cullen, Frank Dimick, and Thomas Knott and produced $5 million in ore over the next eighty years. It produced gold and silver in about equal dollar amounts and also considerable copper and lead. From 1909 to 1917 John Wagner operated the Alta, Palmyra, and St. Louis Mines and grossed $849,147 at a profit of $627,268. His big mill, built in the 1920s, proved a failure. In 1936 Alta Mines built another mill but it burned in 1848. Since then there have only been cleanup activities at the mine.

The Bessie Mine was near Alta. When its stamp mill was built in Alta, the owners drove a connecting tunnel out to Alta from the Bessie. In its time the Bessie Mill was the

The view of the Gold King Mill about 1935 showed just how many trees were cut for cribbing, timbers and firewood. Courtesy Denver Public Library, Western History Dept. (X62342)

largest in Western Colorado, but it used a grinding process instead of stamps. A forty-two hundred-foot tram also brought ore from the Bessie, but the grinding mill couldn't process the Bessie ore profitably, so they went back to using stamps. All processed ore was then shipped to the Ophir Loop by tram.

Back on the paved highway at New Ophir you can take another route to the east to go to Old Ophir and then over Ophir Pass. The road was first opened as a toll road in 1881. The charge was two dollars per pack animal, fifty cents for a saddle animal, and twenty-five cents for a burro. Before that time it was basically a rough Indian trail that went over the pass. Even though the pass reaches 11,789 feet in elevation through a jumble of scree rock, it received much early use by trappers, explorers, and later prospectors traveling between San Miguel Park and Baker's Park. After the RGS arrived, the pass fell into disuse and was eventually closed to traffic. It was reopened in the early 1950s when jeeping became popular. It has been only slightly improved over the years, but it is one of the easiest jeep roads in the San Juans—occasionally traveled in its entirety by owners who don't care what damage they might do to their car. It takes a little over an hour before the road connects with the Million Dollar Highway near Silverton.

Ophir Pass generally follows Howard's Fork of the San Miguel River. George Howard, who prospected all over the Ophir area in 1860 and 1861, ended up settling in Baker's Park and founding Howardsville. However, he often returned to the Ophir area.

Earnest Ingersoll called Ophir "one of the most scenic routes I know of in Colorado.... Stand where you will on the eastern side of the divide between the Animas and the San Miguel, and these great, smooth, cushiony hills of red, tower up level with your eye, burning under the sunlight." He went on to comment on the beautiful tones of orange and red, the deep greens and blacks of the timber, the flowers and lichen, and vast jumbles of fallen rock on the west side of the divide. The scene is just as beautiful today.

Ophir Pass is perhaps most famous for the story of an early-day mail carrier whose route ran from Silverton to Ophir. At the time mail carriers were having trouble crossing the pass. One man nearly froze to death and quit, another found the altitude too frightening and quit after one day's work, and another stole money from the mail and took off to parts unknown. Young Swede Nielson left to make the run on December 23, 1883. He had noticed that it was snowing hard and he knew that avalanches would be running on the pass. But he also felt an obligation to get the mail through from Silverton to Ophir, because he knew that much of his mail pouch contained Christmas pres-

The Alta Mine, Mill and shaft house about 1912. Silver Mountain is in the background. Courtesy Denver Public Library, Western History Dept. Walker Art Studio. (X62364)

ents and gifts of Christmas money. He left in a blizzard carrying sixty pounds of mail on his back and traveling on his long wooden skis. When he didn't show up at Ophir many thought that he had skipped with the money and presents. That fear seemed to be confirmed when several years went by without his whereabouts being known. However his body was discovered by his brother (who had never stopped looking for him) in August about two and a half years after his disappearance in the remains of a huge snow slide with the mail still strapped onto his back. Swede had given his life in an attempt to get the Christmas mail through to its recipients.

The first rich ore in the Ophir Valley was found at the Osceola, which was near the confluence of Howard's Fork and the Lake Fork of the San Miguel River. Augustine Roberts discovered the Osceola in 1877. He built six arrastras near the mine to help process the ore. One hundred thirteen pounds of ore brought five hundred dollars. Ore went out and over Ophir Pass to Silverton to be milled.

Almost immediately beyond New Ophir the dramatic Ophir Needles are visible to the north. They are a favorite spot for technical climbers. Look closely on any summer day and you might see tiny bodies clinging to the cliffs. The Needles (originally called *Cathedral Spires*) are twenty-eight hundred feet high.

The Tidal Wave Mine was located right next to the Ophir Needles. It was discovered in 1893 and was worked until 1910, but it was never a big producer. A bunkhouse and a few buildings were built at the site.

The Carribeau is across the creek about a mile from the paved road. It was filed on July 26, 1882, but as the ore was rather low-grade it wasn't worked extensively until the arrival of the RGS at Ophir Loop. A thirty-stamp mill was built near the mine in 1895. In 1898 it was reported that it had been the chief producer at Ophir for years. One hundred miners were employed that year, and more men worked at the mill. However, its twenty-stamp mill was processing only twenty dollar per ton ore. It produced mainly silver with its best years in 1904-06, but production continued off and on until 1958. Altogether the Carribeau mined about five hundred thousand dollars in ore.

Although the Tidal Wave was never a big mine, it was well promoted-including this rare interior photo. Main Street Photography. Author's Collection.

A short distance further you will note the Suffolk Mine high to the north (left). It was discovered April 3, 1884. It had good gold and constructed its own forty-stamp mill and cyanide plant in 1890 within the Ophir City limits and connected to its mine by a tram. Later the same mill was made into a tube mill and serviced the Favorite Mine with a forty-two hundred-foot tram. It also served another dozen small mines near Ophir. The Suffolk Mine tunneled all the way into Gold King Basin and shipped some ore from that area. The mill burned in 1937, and today only its concrete foundations are visible. The Badger Tunnel was supposed to access the Suffolk vein at a lower level. It hit a few previously undiscovered veins but never made it to the Suffolk vein.

About a mile further is the town of Ophir. Silver Mountain is on the north and Yellow Mountain on the south. Lookout Mountain is to the east and Sunshine is to the west across the Ilium Valley. Leon Linquist staked the first claim on Yellow Mountain in 1875. "Big Abe" Flemings claim became the town site of Ophir. Leon Linquist, Milton Evans, and M. M. Ross built the first house in the eventual town site in the summer of 1877.

Seventeen men spent that winter at the Osceola, which was producing large amounts of gold. Several arratras were built in the valley to process the gold, and the refined ore was then packed out to Greene's smelter in Silverton. By 1879 four hundred to five hundred men were in the Ophir Valley, but many left when rich discoveries were made in Rico. They were replaced by many others who moved into the area that fall and the next year. The town was called Howard's Fork until 1879. The town site company was formed in November 1880, and Ophir was incorporated in July of 1881. The town really came of age in 1882 when Otto Mears finished his Dallas (near present-day Ridgway) toll road and the same year the Ophir to Silverton Toll Road was finished. Ingersoll reported that "only ten years after its discovery both sides of the gulch are pitted like a pepper-box with prospecting tunnels, and there are perhaps twenty mines shipping ore in profitable quantities." He reported that half a dozen were rich gold properties and that several sales had been made for over fifty thousand dollars for a single claim.

By 1879 there were reported to be as many as four hundred claims in the immediate vicinity of Ophir. The miners were so successful with the arrastras that there were ten by 1879. After the smelter was built in Ames in 1882, the arrastras were no longer used. In 1885 Crofutt reported that Ophir had two general stores, a stamp mill, a hotel, a town hall, stables, and about forty houses. He also reported that "the lodes in the area are numerous and rich." The biggest mines in the area were the Gold King, Osceola, Summit, Alta, Chance, Spar, Montezuma, Parsons, What Cheer, and Nevada. By 1897 Ophir had three sixty-stamp mills and a fourth mill under construction. The town also had a newspaper, although it didn't print on a regular basis. Ophir went through the usual boom and bust cycles (but usually had two hundred to four hundred residents) until 1900 when the population went down to 127 (which might have included a few of the people at New Ophir). Nevertheless Ophir had electricity after 1904 and telephone service in 1908.

Ophir also had the usual number of houses of prostitution. The largest brothel was known as the "House of Many Doors" as it had ten doors but no windows. Fires and avalanches later destroyed much of the town. Although the post office didn't close until 1917, the 1905 census showed Ophir's population at only two. One was a Republican and the other a Democrat so they canceled each other out in elections. In the 1950s the Silver Bell Mining Company was started. It operated the Silver Bell and Carbonero Mines for three years. By then Ophir Pass was reopened, and tourists were discovering the area. New Ophir is now rebounding, populated mainly by people who work in nearby Telluride. When you first enter Ophir you will be in the old part of town. Please drive slowly and look out for children. A half mile further is the new part of town (New Ophir in Old Ophir). The area in between is empty because it is in an avalanche area.

As you leave Ophir and travel through the beautiful aspen forests, the tailings of the Carbonero Mill are seen. The Carbonero Mine was high up Silver Mountain at 12,500 feet. Many of the dumps seen on the mountain at this point are connected with the Carbonero. The mine started shipping ore about 1901. Its ore always contained a high percentage of lead. Ten to twenty men were employed at times at the mine, but until 1924 it had no mill, so mules packed hand-picked ore to Ophir Loop. The new mill had a thirty-three hundred-foot tram from the mine, and the mine's best productivity occurred over the next seven years. Then the mine was worked off and on until 1954. During its life the Carbonero produced about $2.5 million in ore. The roads to the mine are private property and usually gated, so stay on the main road, which climbs through a scree slope, passes Beaver Lake, makes a switchback, and comes to the top of the pass (elevation 11,789 feet). Lookout Peak is directly to the north and South Lookout Peak to the south. Although you would never know it from the road, there is a beautiful little lake to the south of the pass called Crystal Lake. It is well worth the hike.

There are wonderful views from the top of the pass back toward Ophir. Left to right (south to north) the mountains in the distance are Lizard Head (13,113 feet), Mount Wilson (at 14,246 feet the second highest mountain in the San Juans), El Diente (14,159

The town of Ophir included several mills and trams that came right through the residential section. Courtesy Denver Public Library, Western History Dept. Thomas McKee Collection. (X11517)

feet and Spanish for tooth) and Wilson Peak (14,017). Both of the Wilsons were named for A. D. Wilson of the Hayden Survey party.

After crossing the divide you are on San Juan County Road 8. A mile down the road a mine road (usually gated) leaves to the U.S. Treasury. Then another mile down the road the Gem Mine and dump are directly below the road (south). Another mile down the road there is a private road that leads down to the creek and the Bonner Mine. Then a half mile further, just before tying in with the Million Dollar Highway, San Juan County Road 100 leaves to the left. It travels up Mineral Creek for a mile or so and then loops around the Imogene Mine (Andy Richardson at work again) and ties back into the Million Dollar Highway. About two-thirds of the way down this road is the trail to Columbine Lake. It is a steep hike but very beautiful. The lake is under Lookout Peak and Mill Creek flows out of the lake and down to the Chattanooga Valley.

The Ophir Pass Road crosses Mineral Creek at what was called Burro Bridge. There was once a small community here. The name came from a small, narrow bridge across the creek that was adequate for only horses or burros. At exactly ten miles the Ophir road ties in with the Million Dollar Highway.

Back at Ophir on Highway 145, take the paved highway to Trout Lake, Lizard Head Pass, and Rico. Shortly after the climb up the higher side of the Ophir Loop is the site of the San Bernardo Mine. It was located in 1877 but was not worked heavily until the mid-1880s. A mill was built on the site, but it was destroyed in the 1909 flood. Then another mill was built in 1920. The mine went through boom and bust from 1920 through 1969, but by 1945 the San Bernardo and several nearby mines had produced over $5.5 million in ore. The mine was the last operating in the Ophir area, although the Silver Bell Mill was processing ore from a mine on the other side of Ophir Pass until 1976. There was logging nearby and the San Bernardo Mill also processed the ore from a few small mines nearby. The mill still stands and is very complete, but it is private property.

The little town of Matterhorn was built close to the mill. The population of Matterhorn was never big; twenty-seven in 1900 at its peak and only seven in 1940. There were eight or ten houses and a saloon at one time. Matterhorn was originally called San Bernardo (it was named for the nearby mountain to the west) but it was renamed because of confusion with San Bernardino, California. Ironically the post office never changed the name.

The Priest Lakes Recreation Area, which is nearby, is named for a pioneer family that lived in the area. A little further is Trout Lake, which has always been a beautiful natural lake, although it was dammed and expanded to provide water by pipeline to the new power plant at Ames. The Walton Expedition of the St. Louis Fur Company spent a fair amount of time at or near the lake in 1833. Originally, it was called Fish Lake. As early as 1874 Rhoda and the San Juan Division of the Hayden Survey commented that the lake was filled with "fine trout." The expedition named it San Miguel Lake but the fish won out. Even in the 1870s the miners could in one day catch hundreds of trout from the lake. A town site was platted for Trout Lake but never built. A few cabins were located on the north end of the lake near the east end of the dam. In 1883 a post office and seven cabins surrounded the lake. The Telluride Power Company leased sites around the lake to individuals and many small cabins were built. The RGS had a fifty thousand gallon tank at the site. It is still standing—a favorite photo stop for rail fans. There is another natural lake above Trout Lake called Lake Hope, and the power company built an artificial lake in between called Middle Lake. Trout Lake covers 138 acres, Middle Lake is 50 acres, and Lake Hope is 38 acres.

From the top of the divide (10,222 feet) one can see Lizard Head (13,113 feet). The rock does look a little like a lizard looking up towards the sky. It has always been a favorite subject for photographers and was used on the logo for the Rio Grande Southern Railroad. Albert Ellingwood and Barton Hoag were the first to climb Lizard Head in 1920. Because of its steepness and "rotten' rock, it is considered to be the hardest thirteen thousand foot peak to climb in Colorado. Rhoda explored Lizard Head Pass in 1874 and noted that it was "marked with a curious monument of trachyte two hundred and ninety feet high." He also climbed Mount Wilson and Mount Sneffles while in the area. The RGS sometimes needed four or five engines to clear the snow over Lizard Head. Otto Mears even built a quarter-mile long snow shed to cover a wye at the top of the pass, which was the highest point on the RGS line. A section foreman always lived at the site, and besides his house there was a bunkhouse and livestock corrals. A post office established at Lizard Head on two occasions, but abandoned each time after only a few months. An earthquake caused a huge landslide in the early 1900s, which changed the looks of the peak when millions of tons of rock fell off. It shortened the summit by forty-six feet.

As you come to the south end of Lizard Head Pass, the road follows the Dolores River (originally called Rio Nuestra Senora Delores which means "River of Our Lady of

Sorrows" in Spanish). The name was officially established by the Escalante-Dominguez expedition. At several locations the old RGS right of way can be seen. The Lizard Head region is so remote that when the RGS official photographer, W. H. Jackson, came to the pass in the mid-1890s, it was the first time that the region had ever been photographed.

The Coke Ovens depot and siding was a few miles further at the junction with the Dunton road. The mining town of Dunton is ten miles down the road. It has a wonderful hot springs resort. There were charcoal kilns at Coke Ovens (hence the name). Coke is produced when volatile components of bituminous coal are driven off by heat and the carbon and ash components are driven together. Coke was used in smelting because of the intense heat it produces when burned. The depot served a few nearby mines as well as Dunton. Very little remains today, but at one time it had a water tank, several houses, and a siding. The Emma, Smuggler, and American Mines were about a mile before the town. Dunton was established in 1882 as a mining town called "Bowen," but it was a short-lived boom. Then in 1883 Horatio Dunbar built cabins, a hotel, a store, and a saloon. Its population peaked at about 280 people in 1905. By 1918 the town was basically deserted but a few people have always resided at Dunton year-round because of the hot springs.

The main road to Rico has a very narrow section in Burns Canyon. It was so narrow that the RGS basically built a bridge that went straight down the river, and then its track continued on the steep bank on the other side. A short distance further is the town of Rico. As you come into town from the north, Silver Creek divides Nigger Baby Hill behind the town on the north from Dolores Mountain on the south. Newman Hill forms a bench-like extension of Dolores Mountain. Many of the rich mines in the area lie on these mountains or up Silver Creek.

Rico was known from the earliest times to have valuable ore because it was on a branch of the Old Spanish Trail. Rivera passed this way in 1765, Escalante and Dominguez in 1776, and there were enough unofficial prospectors that when Captain William Walton and sixty fur trappers of the St. Louis Fur Company passed through in 1832-33 they noted the remains of several old Spanish smelters. In 1861 George Howard led a group of prospectors up the Dolores on their way to what is now known as Ophir.

Americans also prospected the area in 1866 when a Colonel Nash led eighteen prospectors into the vicinity of present-day Rico. In July 1869 Shelly Shafer, Joseph Fearheiler, and seven other prospectors were in the area and discovered the Pioneer and the Nigger Baby lodes. In 1870 R. C. Darling ran across Shaefer and Fearheiler at the area, built a cabin on Silver Creek, and the group located several claims including what eventually would become the first big mine—The Atlantic Cable. In 1870 Gus Begole, John Eckles, Dempsey Reese, and "Pony" Whittemore also prospected around Rico and discovered the Aztec and Yellow Jacket Mines. Darling left but returned with other men in 1870 and even built a crude furnace to help smelt the ore. Unfortunately the land still belonged to the Utes so no legal claim could be filed, the ore turned out to low-grade, and transportation was a major problem. What ore was taken out went over the pass to Rockwood north of present-day Durango. The freighters followed remnants of a trail used by Native Americans for centuries before the whites arrived, but the easier route from Silverton was over Ophir Pass, then Lizard Head Pass, and down the Dolores River.

In 1874 the Hayden Survey party passed through Rico. Despite all the activity, the area (located at 8,737 feet) was very sparsely populated until richer discoveries of gold and silver were found in 1878 by John Glascow and Sandy Campbell on the Grand View and the Atlantic Cable and in 1879 when Col. J. G. Haggerty found rich ore at the Phoenix Mine, nicknamed "The Nigger Baby." The hill became known for the mine, which was rich but was also the focus of many scams. Sidney Jocknick wrote that "so many claims were salted to catch tenderfeet that its bed would have stood tests as relics of Sodom." The Puzzle Extension, for example, was salted and sold three times until a genuine bonanza was discovered. By August of 1879 the town of "Carbonate Camp" (evi-

Louis McClure shot this photo of Lizard Head in the dead of winter about 1910. Courtesy Denver Public Library, Western History Dept. (MCC2681)

dently someone was thinking of another Leadville) contained 105 private dwellings (many of them tents) and twenty-nine commercial buildings, which included seven saloons, four assay offices, a sawmill, newspaper, church, and a post office. The first big sale of claims was made in August of that year for one hundred thousand dollars and included the Phoenix, Yellow Jacket, Pelican, Aztec, and others. The first woman came to Rico in August 1879, but soon thereafter the first murder occurred—George McGoldrick, alias "The Kid" shot and killed "Frenchy," who became the first burial at the new cemetery. That winter there were eight women who braved life in the isolated town. After considering at least ten other names, the name "Carbonate Camp" was changed to Rico in 1879. Rico means rich in Spanish. The *Dolores News* was operating by August 1879, the first issues being printed in Silverton, but by October there was a press at Rico. In 1879 a fire department, militia, and literary society were formed. There was no real organization to the mass of people there, which made the Ouray paper note "this place impresses one as having gotten there before it was sent for."

Rico, like all of the San Juans, had its Indian scare in 1879 after the Meeker Massacre. On the night of October 5, 1879, the citizens of Rico went through a major fright when they received news of the Meeker Massacre. The women were taken to an unfinished log cabin with no windows or doors. The men took refuge in Frank Raymond's store. When a burro puncher showed up, a nervous guard shot one of the burros, whereupon the freighter beat up the guard. While all this excitement was going on, Robert Spencer on October 11. 1879, became the first white child born in the Dolores Valley. Rico was officially incorporated in December 1879, and the town was officially surveyed and platted in 1880.

The spring of 1880 saw rapid growth, and the town's promoters estimated as many as three hundred homes and one hundred other structures had been built. The population was estimated at twelve hundred. This was probably an enormous overstatement, but many new businesses were opened in town including six general stores, three blacksmiths, a school, and four butcher shops. It took sixty-six days to bring the machinery for a smelter by wagon from the end of the railroad at Alamosa, but the smelter was producing five months later. The Grand View Smelter was also built that year and officially "blown in" on November 7, 1880. Rico had the same problems as most of the San Juan towns—lack of capital and high transportation costs. As the rich ore on the surface was found to be lower-grade with depth, the population of the town again dwindled.

In 1881, after the Utes had been forced out of Colorado, Dolores County was carved out of Ouray County, and Rico was made the county seat. In the spring of 1882, the Newman group of mines were purchased by Marrs Consolidated Mining Co. for $175,000, and that same year the trail up Scotch Creek was made into a good wagon road and two stages daily came over the pass from Durango to Rico. Generally, oxen were used to move freight in and out of Rico, which indicated that the roads were wide and clear enough to be a wagon road but that they were steep—as many as twelve oxen pulled the huge freight and ore wagons. Today this road comes out at the Durango Mountain Resort ski area. In the 1880s the road began at Rockwood near current-day Tamarron, went into the mountains to the west at the ski area, then went through Hermosa Park, and finally over the range to the Dolores drainage. However, it was a long and rough road, and one writer in 1881 wrote, "Rico, when you get there, doesn't look like too much."

In June 1882, Tom Well and "Trinidad" shot and killed City Marshall Smith who was trying to arrest them for stealing a saddle. When he fell to the ground, they emptied their guns into him and then fled to the south and the Ute Indian Reservation. White men on the reservation turned them in for the reward, and they were brought back to Rico. In the night the guard was jumped and the men lynched. The bodies were left hanging as a warning to the "bad" element in town.

The winter of 1883-84 was so rough at Rico that all the residents pooled their food and then split it, depending on the size of each family. The snow was reported to be eight feet deep on the level. Several residents were found hiding food from the others and their supplies were confiscated. A man bringing relief supplies was held up just a half mile from town and the supplies taken. An old ox that somehow survived the winter was butchered and sold for a dollar a pound.

Ingersoll visited Rico in 1884 and declared:

The town itself is made up of a scattered, gardenless collection of log cabins and some frame buildings with a log suburb called Tenderfoot Town, and numbers about six hundred people. It is very dull compared to most Colorado camps, but this is owing to the fact that everybody is waiting until the railroad gets a little nearer..... Rico has not yet been worked far enough into her numerous 'locations' to make sure of the riches her mountains are suppose to contain.

Oxen were the preferred method of pulling the heavy freight wagons to Rico in 1890. Courtesy Denver Public Library, Western History Dept. (X13156)

At the same time Crofutt stated that Rico was thirty-five miles from Silverton as the crow flies and had a population of fifteen hundred. "Near Rico have been found some of the most prolific, easily mined, precious metal veins yet discovered....Rico is to a great extent self-sustaining, as nearby are found coal, iron, lead, salt, timber, fire-clay, and in fact everything necessary for cheaply producing bullion by milling or smelting process." But silver prices started sliding and by 1885 the census only listed 894 residents and then only 400 in 1886.

Tradition has it that in 1885 one of the local saloonkeepers by the name of David Swickhimer ran out of money to operate a mine that he had been working since 1881. His wife, Laura, who ran a boardinghouse and restaurant, also poured every extra cent she had into the mine. When she won five thousand dollars on an eastern lottery ticket she invested the money into the mine against her husband's wishes, but the crew struck a rich vein on October 6, 1887. The ore ran four hundred dollars per ton in silver and gold, and there was a lot of it! By the end of 1887 the mine—the Enterprise—was known to be one of the richest in the San Juans. Ironically, David and his wife divorced after the big strike. The mine was eventually sold in 1891 for $1.25 million; however, David bought the Rico State Bank and when the bank failed in 1907 he spent a large part of his fortune reimbursing depositors.

The 1890 census showed 1,134 people living in Rico and another 400 just outside the city limits but living close to town. The Congregational Church was started that year by the Rev. T. C. Dickerson. The town boomed again when on September 30, 1891, the Rio Grande Southern Railroad construction train arrived. Originally the railroad was supposed to arrive by November 1, 1890, if it was to receive free land for the depot and track through the town, but the time was extended. Formal service started October 15,

221

Many of Rico's large mines were on the hills just behind the town. Buckwalter Photo. Courtesy Colorado Historical Society. (CHS-B169)

The Enterprise was one of Rico's really big mines. Courtesy Denver Public Library, Western History Dept. (X13149)

1891. The railroad, which ran from Ridgway to Durango, was started simultaneously from both ends and met at the middle near Rico. An engraved silver spike was driven to commemorate the completion. There were days of great celebration when the first train arrived. Freight rates fell from fifty dollars a ton to eight dollars a ton overnight. Not only could the ore be shipped a lot cheaper but it meant the supplies coming in to Rico were much more affordable.

Many of the Rico mines reopened and the population swelled. In 1892 Rico was at its peak. There were eighteen hundred miners at work and almost five thousand people at or near the camp. Twenty-eight saloons furnished liquor and sex to the hungry miners. But the prosperity was short-lived because the Silver Panic of 1893 again closed most of the mines. Rico's peak came right before the Silver Panic, yet its mines still managed to produce over two million dollars of ore in 1893. After the Silver Panic, fortunes were lost overnight, and most of the businesses in the town folded. Even the railroad went bankrupt, although it was taken over by the D&RG. Things did get better, and Rico's fortunes again peaked in the early 1900s when large amounts of tin were being mined. Metal prices rose during World War I, but they were generally in a downward spiral. The population of Rico was 364 in 1910 and under 150 in 1920, then things rebounded in the late 1920s only to fall again during the 1930s. Then in the 1940s things picked up again under Robert and Elizabeth Pellet, who leased some of the better mines. The population rose again to almost twenty five hundred. The Rico mines had produced over $15 million in ore by 1920, but it was a small amount compared to Ouray, San Juan, and San Miguel Counties. By 1945 production had risen to almost $27 million, but mining busted again in the early 1950s.

It was the area's mines that controlled Rico's fate, but their production was eighty-eight percent silver, and after 1893 the San Juan mines were looking for gold. Altogether the Rico mines produced about a half billion dollars in today's values—certainly it had earned the name "rich." Large mines in the area included the Yellow Jacket, Black Hawk, The Iron Mine, Argentine, Grand View, Jumbo, Enterprise, Montezuma, Newman, Telegraph, Aspen, Grand Prize, and Albany—all generally close to town.

Some of the old buildings still standing today include the two-story, red-stone Burley Block built in 1891 (the fire department is just next door), the headframe for the shaft of the Atlantic Cable Mine, the Old Masonic Hall, the Enterprise bar (the old Dey building), the Presbyterian Church (previously the Congregational), and Town Hall (the original County Courthouse built in 1893 by C. A. Ransome). The Dolores County Seat was moved to Dove Creek in 1944 because Rico's population was so small. Now the situation is reversing itself, as Rico regains some of its former glory because of the boom in nearby Telluride.

OURAY

Camp Bird Rd.

Amphitheatre Campground

Uncompahgre River

U.S. 550

Tunnel

Hayden Trail

Bear Creek Rd.

Bear Creek Falls

Silver Point Mine

Union Mine

Yellow Jacket

American Flats

Horsethief Trail

Sutton Mine

Sivyer Mine

Grizzley Bear Mine

Dunmore Mine

Engineer Rd.

Riverside Slides

Uncompahgre River

To Engineer Pass

To Cinnamon Pass

Hayden Mountain

Crystal Lake

Abrams

Lucky Twenty Mine

Hendrick Gulch

Brown Mtn.

Ironton Park

Richmond Trail

Tailings pond

Ironton

Red Mtn. 1

To Cinnamon Pass

Mtn. King Mine

Trico Peak

TELLURIDE

Black Bear Mine

Bridal Veil Falls

Barstow Mine

Black Bear Rd.

Meldrum Tunnel

Joker Tunnel

Silver Bell

Corkscrew Turntable

Red Mtn. 2

Guston

Yankee Girl

Idarado National Belle

Genessee Vanderbilt Mine

Longfellow

Red Mtn. 3

Lookout Peak

Summit

Congress

Red Mtn. Town

Red Mtn. City

Bullion King

Silver Ledge

St. Paul

Porphyry Basin

Chattanooga

Brooklyn

N

W   E

S

Burro Bridge

Ophir Rd.

Clear Lake

Ice Lake

Christ of the Mines

Walsh Smelter

SILVERTON

Bear Mtn.

North Star Mill

South Mineral Campground

U.S. 550

Bandora

To Durango

Chapter Eleven

RED MOUNTAIN &
THE MILLION DOLLAR HIGHWAY

There is probably no drive on a paved highway in the state of Colorado offering more awesome and spectacular scenery for so many continuous miles as the Million Dollar Highway, which runs from Silverton to Ouray. Because of its narrow, curving design, it is not a road to be traveled fast (the speed limit is 25 mph on most of the route anyway), but that can be an inspirational advantage. There are numerous places to pull off along the route, where you can stop and enjoy the views, but if you drive the highway be sure not to stop in the marked avalanche areas during the winter—as the Million Dollar Highway is one of the most avalanche-prone highways in the United States; but you will only be in a typical avalanche path for a second or two, which makes it very unlikely that you will be hit—unless you stop!

Many people who travel the Million Dollar Highway for the first time wonder how it got its name. Some speculate it originally cost a million dollars to build, and some estimate it would cost a million dollars a mile to build today. Others say that the dumps from the local mines were used for road base and they contained a million dollars in unmilled gold. Some feel there is a million dollars worth of scenery along the route. Another story is that the name came from a woman who traveled the road and said, "I wouldn't go back over that road for a million dollars!"

None of these colorful tales is correct. The name came into use between 1921 and 1924 when the road was being reconstructed for automobile travel. By far the most difficult part of the project was the twelve-mile section from Ouray to Red Mountain Pass, and that portion was further divided into three segments with contracts given to different contractors. In a planning meeting, one of the contractors mentally summed up the bids and was struck by the fact that the total was almost exactly a million dollars—a large sum for 1922! The phrase caught on and was in general use by the dedication of the new highway two years later. The Million Dollar Highway of 1924 was not paved and was basically only one lane wide (with turnouts for meeting and passing oncoming traffic) but it was as up-to-date and auto-worthy as any mountain pass in Colorado at the time.

It is often written that Otto Mears built the original Million Dollar Highway. Actually it wasn't quite that simple. On April 1, 1880, the Ouray and San Juan Wagon Road Company filed its articles of incorporation. Its purpose was to build a wagon road from Ouray to Mineral City (Mineral Point) with a branch running up Poughkeepsie Gulch and a branch running up Red Mountain Creek into Red Mountain Park (now called Ironton Park). At this point Otto Mears was not involved. Although some progress was made between 1880 and 1882, it was extremely slow, and in the spring of 1883, it appeared that the Ouray and San Juan Wagon Road Company was in deep trouble. At the same time there was a great need for the branch road to Red Mountain to be completed, because the Red Mountain Mining District was booming.

Early day tourists enjoy the Million Dollar Highway by stagecoach. Postcard in Author's Collection.

To speed up the process, Ouray County undertook to build the portion of the road from the Uncompahgre crossing (current Engineer Road) to Ironton Park as a public road.

After the county had taken over the branch line to Red Mountain, the toll road company was able to concentrate its efforts on the construction of the main road. The county made considerable progress toward Red Mountain during the early weeks of the summer of 1883, but problems with financing still plagued the original builders and their part of the road. It was at this point that Otto Mears concluded a deal with the Ouray and San Juan Wagon Road Company, whereby he accepted twenty-seven thousand dollars of unissued stock (which gave him fifty-four percent ownership) and agreed to complete the building of the toll road during that season. He put on a much larger work force and by the first week in October, eight- and ten-mule outfits could travel the road as far as Poughkeepsie Gulch. By the next spring the road was completed by the county to Red Mountain, and one year later Mears completed a toll road from Silverton to Red Mountain to complete the route.

To travel the current Million Dollar Highway, beginning at Ouray, follow the paved highway out of town up the switchbacks to the south. The Camp Bird Road is the first road to the right, then the road to the Amphitheater Campground (open from Memorial Day to Labor Day) leads to the left. A little more than a mile further, the road passes through a tunnel originally built by Lars Pilkaer in 1921. The tunnel was enlarged and reinforced during 1998 and 1999.

Immediately on the other side (south) of the tunnel is the beginning of Bear Creek hiking trail, which is exciting and well maintained. It takes two to three hours to reach the Grizzly Bear Mine and several more hours to get to American Flats and Engineer Mountain. The trip can even be made into a two-day hike by combining with Horsethief Trail. The trail was originally on the south side of Bear Creek and was one of the principal early day routes on horseback or by foot from Ouray to Lake City. It is still possible to travel the south trail but it is not easy. Hiking the north trail, the

Sivyer, the Silver Point, Painted Chief, Union, and then the Silver Queen mines are visible to the south.

The first mine with improvements on the north side of the creek is the Grizzly Bear Mine, which was located June 16, 1875, by L. W. Balch and F. W. Sitterly and reached peak production while owned by George and Ed Wright and Milton Moore. When George Hurlburt bought the Grizzly Bear in 1896, he switched the original trail to the present trail to the north side of the canyon so that the snow and ice wouldn't be such a problem. A small town was located at the Grizzly Bear Mine, with the 1900 census listing a population of twenty-four. Unfortunately, most of the extremely rich ore was exhausted by that time, and the inaccessibility of the mine made it impossible to mine the lower grade ore. The Grizzly Bear

This early photo shows the Bear Creek trail was carved out of the cliffs at some points. Author's Collection.

Mine itself is located across the creek from the few remaining buildings, and a bridge used to run across the canyon from the boardinghouse to the mine. Today the mine is worked sporadically, although it is now accessed from a tunnel in the amphitheater above Ouray.

Continuing on the trail for about two miles east you will hike the steep "Hell's Half Mile," then gently travel to an elevation of eleven thousand feet and the Yellow Jacket Mine. The Yellow Jacket was most active around 1915 when a small mill was built and machinery hauled in—even though no large amount of ore had been discovered. Then the whole operation was abandoned! From this point the trail forks. The northeast trail leads to Horsethief Trail, while the southeast trail goes to American Flats and eventually ties in with the Engineer Road at 12,700 feet near the top of Engineer Mountain.

Somewhere alongside Bear Creek is the location of the "Crazy Swede" Mine. Gus Lindstrom found an extremely rich vein while coming down Bear Creek in a snowstorm in 1906. Ore from the one-foot-wide vein supposedly assayed at $350,000 a ton, but the Swede could never find the vein again, although he searched for it for decades. (See *Sheepherder's Gold* by Temple Cornelius.)

Back on Highway 550 and traveling south about a tenth of a mile from the tunnel, one comes to the Bear Creek Falls turnout. The thin, trickling waterfall formed by

The Million Dollar Highway and Bear Creek converged at Otto Mears'
tollgate. Author's Collection.

Ralston Creek across the canyon is called by several names—"Horsetail Falls" or "Indian Tears" being the most common. Roger Henn in *Lies, Legends and Lore of the San Juans* reports that the latter name came from a battle between the Utes and another tribe. The intruding tribe was being slaughtered in the canyon below and the waterfall started from the tears of their women and children who were watching from above.

At one time a huge glacier, probably three thousand feet thick, stretched for twelve miles down the valley (north) to Ridgway. It was this glacier that formed the canyon not the river (which probably has cut into the earth by no more than fifty feet). Bear Creek drops 227 feet into the Uncompahgre River. In the winter it is one beautiful blue-green icicle—a favorite spot for the local ice climbers.

The tollgate to Otto Mears's road stood right on top of Bear Creek Falls. There was no way to get around the toll because of the cliffs and the steep drop-offs. A twelve-foot-wide bridge of logs crossed the top of the falls. Toll was at first $3.75 for a wagon with two animals, $.75 for a horse and $.35 for each pack animal. When the toll road charter expired in 1900, the county took over the road and made it public.

Across Bear Creek Falls to the south was the Sutton Mill. It was built by Jim and Bill Slick in 1926 and had a one hundred-ton capacity. The mill was mostly destroyed by fire in 1982. The Sutton Mine is across the canyon at an altitude of 10,595 feet and was connected to the mill by a twenty-seven hundred-foot aerial tramway. Another access to the mine was from the Mineral Farm area off the Camp Bird Road. The mine had four working levels with most work being done in the 1920s out of the Barber Tunnel, which was connected to the mill. The mine produced pyrite, lead, copper, silver, and some gold. Usually one of the mine's cabins has clothes hanging outside and a sign proclaiming "Antiques." It is simply a ruse by locals that usually brings a chuckle from observant visitors.

Immediately alongside the road at Bear Creek Falls is a monument to Otto Mears who truly deserves the title "Path Builder of the San Juans." In 1887 Mears began construction on the Silverton Railroad that ran from that town to Red Mountain. In 1889 he began the Rio Grande Southern Railroad, which ran from Durango to Telluride and then over Dallas Divide to Ridgway. He had a flair for the unusual and issued silver and gold passes to his friends for use on the railroads. Mears became a powerful political figure in Colorado. He helped plan the state capitol and many other state buildings. At age seventy-seven he retired and moved to California where he died on June 25, 1931. His ashes were scattered at Bear Creek Falls. The monument was erected in 1926. See *Otto Mears* by Eva Tucker for a good account of Mears' life.

Approximately a mile above Bear Creek is the four-wheel-drive jeep road to Lake City over Engineer Pass. The Uncompahgre River passes under Highway 550 at this point. The stream that follows the paved highway is Red Mountain Creek. About a quarter of a mile south of the Engineer Road, U.S. 550 passes beneath the Mother Cline Slide. The slide takes its name from the Mother Cline Mine, located above the highway. It was one of the earliest mining claims filed in Ouray County, having been located on August 14, 1875 by Milton W. Cline, one of the founders of Ouray and the town's first mayor. The Clines had a cabin on the hill near the main highway. Travelers and miners would often stop at the cabin for food or to warm themselves.

The Riverside Slide has always piled up deep in the Uncompahgre Canyon.
Author's Collection.

About four-tenths of a mile past the Mother Cline, on the other side of the canyon, is Silver Gulch. The Dunmore Mine is located near the bottom of the gulch. The original mine was worked for many years and produced tungsten as well as other minerals. In the 1930s G. A. Franz bought the mine and built a tram from the mine to the highway.

Three-quarters of a mile further south is the infamous Riverside Slide. The slide takes its name from a mine (now buried) that was located right on the river. In earlier days the road was much closer to the creek and quite often a snow tunnel was necessary in the spring. The slide comes down both sides of the canyon and sometimes covered the old road fifty to seventy-five feet deep for 450 feet or more. In 1985 the present concrete snowshed was completed, which took some of the risk out of driving the highway in the winter, but it was just a portion of the snowshed that had been originally proposed. Not building the snowshed was to prove to be a tragic mistake.

Many people have been killed at the Riverside Slide. In 1897 the mail carrier, John "Jack" Bell, was buried but dug himself out after twenty-four hours. In December 1908, the slide took its first life, Elias Fritz, who was riding in a freighting sleigh coming down from Red Mountain. Several near fatalities occurred, including a near miss of the stage in 1910. On March 3, 1963, the slide dealt death again, killing the Rev. Marvin Hudson and his two daughters, Amelia and Pauline. About a quarter of a mile south above the snowshed, a monument has been erected to remember the tragedy. The family was traveling to Silverton for church services when the tires on Rev. Hudson's car began slipping and he stopped to put on chains. He had only a few seconds warning before the slide hit. It was a week before Rev. Hudson's body was found. Amelia's body was not found until March 17, and Pauline was not found until May 30 of that year. The car was found downstream six hundred feet from where it was swept off the road with the top torn off and the doors either open or gone. However, a jar of cream in the car was unbroken.

Next to the Hudson monument is another memorial honoring three snowplow drivers who were killed at different times while clearing the slide. On March 2, 1970, Bob Miller died while using a Caterpillar tractor. On February 10, 1978, the slide killed Terry Kishbaugh who was driving a rotary plow. The name of Eddie Emil was added after he was killed standing just outside his plow near the snowshed on March 5, 1992. If the snowshed had been its original length Eddie Emil would have lived.

Traveling up the road about half a mile, U.S. 550 enters a large flat area called Ironton Flats or Ironton Park. The Red Mountains are visible at the southern end. The western ridge of the valley divides the Red Mountain District from the Imogene District. The eastern ridge is known as Brown Mountain with its northern extremity being Mount Abrams. The mountain has two names! Brown Mountain is 13,339 feet at its south end, and 12,801 feet at Abrams on the north. The flats were formed when large mudslides from Hendrick Gulch (which enters from the east) dammed the lower end of the valley forming a lake that gradually filled with silt. Good "flatland" cross-country skiing is available in this area in the winter.

The Lucky Twenty Mine (also called the York and the Guadeloupe after various tunnels that progress up the mountain) is located about halfway up Hendrick Gulch. T. J. York, postmaster and county clerk in Ouray in the 1890s, originally owned it.

This shot (taken about 1888) of Ironton even included a man and his bicycle. Author's Collection.

The mine has operated off and on to present times. It averaged about fifty ounces of silver and twenty ounces of copper per ton of ore and veins up to twelve feet wide were found.

The north end of Ironton Park was developed in the late 1930s as a year-round recreation area. Ralph Kullerstrand and Joseph Condotti constructed the dam that formed the lake (now called Crystal Lake) and built a ski lift on the east side of the road with a run of approximately eighteen hundred feet. They razed the giant smokestack at the Saratoga Smelter and used the brick to build a lodge. Unfortunately, a disagreement arose between the two partners, and the resort was never opened for business. Members of the St. Germain Foundation ("I Am" religion) purchased the property in the 1940s but occupied it only in the summer months. One day in the winter, the watchman was attempting to melt a small patch of ice on the roof with a blowtorch, which caught the roof on fire and the lodge building burned to the ground! All that remains today is the stone garage and basement.

By following the trail to the north of Crystal Lake, one can get to the beginning of the Hayden Trail, which is approximately six miles long. In normal years its entire route can be traveled safely only in July and August. It can be completed in about seven to eight hours but requires very strenuous, steep climbing. The entire trail doesn't appear on current topographical maps, but the forest service and volunteers have cleared and marked the trail. It ends at the Camp Bird Road near the spot where the road crosses Canyon Creek.

Traveling south through Ironton Park (a spot that most of the locals believe to be the best in the area for fall colors), several small mines are visible. On the east side of the park, from north to south, are Albany Gulch, Brooklyn Gulch, and Avarado Gulch, each of which once contained a mine by the same name. Albany Gulch also contains the ruins of the Saratoga Smelter and Mine, which was the northern terminus of the Silverton Railroad. The mine was large and contained good ore, but by the mid-1890s transportation costs were too high, despite the presence of the railroad, and the price of silver was too low, so the mine was abandoned. The Saratoga was discovered in 1883, and by 1894 had produced $125,000, mainly in silver with some gold. It eventually had a one thousand-foot tunnel with several levels.

About a mile and three-quarters into Ironton Park is the beginning of the tailings from the Idarado Mine. The Corkscrew Road (covered in another chapter) begins at this point. The tailings cover a large part of the Ironton town site, and the area has been revegetated. The spot where the tailings pipeline crosses Corkscrew Gulch is the second longest suspension bridge in Colorado (the Royal Gorge is first). The pipe was made to bow upwards when empty so it would be level when filled. The Larson Brothers (or Ida L.) Mine is visible on the other side of Highway 550. The mine had two veins that have been explored for about 180 to 200 feet each. The vein averaged about five feet wide, and production consisted mainly of silver with very small amounts of gold, zinc, and lead.

The Larson brothers (Harry and Milton) were the last residents of Ironton. They preferred to live in Ironton when every one else had moved away. After Harry died Milton gained considerable notoriety when he appeared on "What's My Line," as the only resident and "mayor" of Ironton. He died in June of 1964, at which time his house was boarded up, and Ironton officially became a ghost town.

A road follows the creek a little more than a half-mile south alongside the tailings pond to what is left of Ironton. The settlement was founded in 1883 and platted March 20, 1884. It was originally called "Copper Glen." By the end of March, thirty-two cabins were already being built. Less than a month after its founding, Dave Day announced in *The Solid Muldoon*, "Three weeks ago there was not a building to be seen on the spot where Ironton now stands. There are now over one hundred

This early photo at the Beaver Mine was obviously posed, as the men are not in their working clothes. Courtesy Denver Public Library, Western History Dept. (F3091)

buildings of various dimensions under headway, and the inevitable dance hall in full blast." The main problem was that lumber could not be brought to the new town fast enough to keep up with the building that was taking place. Sawmills were set up near the town, but the trees disappeared fast; they were not only being sawn for lumber but also used whole for log cabins, cribbing, mine timbers, fuel for boilers, and heat. It didn't take long for the nearby slopes to be denuded. The name of the settlement was changed to Ironton in April, and Ironton got a post office in May. By mid-1884 Ironton was the biggest of the Red Mountain towns. It had twelve saloons, four restaurants, and even a newsstand that also sold fine cigars. It had only one main street but it was almost a mile long! It was the supply town for many rich mines in the area and later in 1889 the passenger terminus for the railroad. By 1889 it had a water works and electric power plant. Dave Day wrote: "Ironton's residents now bathe once a month.... As a beverage they have no use for water." The 1890 census listed a population of 323, and the town had two churches and a fire department. With the arrival of the railroad the town saw a real boom, especially in the summer when as many as two hundred "Circle Route" tourists might be in town on any given day. In the late 1890s the area began to wane with the fall of silver prices, but the subsequent discovery of small amounts of gold kept the area alive. Today, only a half dozen houses still stand as most were torn down for their lumber or covered by the tailings, but Ironton is still one of the best-preserved ghost towns in the San Juans.

A short way further up Highway 550, a small road to the west side of the highway leads to the Beaver and Belfast Mine. The mine was an early producer and contained several thousand feet of tunnels following one-foot to four-foot wide veins of gold, silver, lead, and zinc. Shortly past this mine, the Richmond Trailhead leads up to the Greyhound Mine and on over Richmond Pass into the Camp Bird area. The trail is not maintained and the road passes through private property, but is a very interesting eight-to-ten hour hike. The mine was located to follow an extension of the Camp Bird's rich veins and has been worked off and on up to the present.

Further along Highway 550 you start into the heart of the Red Mountain Mining District, which is explained in detail in P. David Smith's *Mountains of Silver*. Red Mountain is actually three different mountains. The highly oxidized iron in the mountains makes them appear to glow after rains. However, it was something different the prospectors were after. The area contains a number of pipe-like mineralized areas of unbelievably rich silver-lead and silver-copper ores.

The Red Mountain Mining District was first prospected in September 1879, but the area didn't really take off until 1882. Gold, silver, lead, and copper were produced. The *Red Mountain Pilot* of April 28, 1883 reported that there were:

...five towns in the district as follows: Chattanooga, eight miles from Silverton and D & RG Railroad; Red Mountain City, one mile distant (both of these towns are in San Juan County, the latter being about half a mile from the county line); Red Mountain, or Hudson Town is one mile from Red Mountain City, in Ouray County and half a mile from the county line; Rogerville is about half a mile from said Hudson Town; and Ironton is four miles distant from Hudson....

Traveling south on Highway 550, the Silver Belle Mine is visible across the creek. It operated from 1880 to 1894, but became the first of the Red Mountain mines to have to shut down because of high water, sulfuric acid, and the prohibitive costs of the deep shafts following the volcanic pipes. It was one of the best producers in the area with ten levels and select ore running up to four thousand ounces of silver per ton. It had an output of over a million dollars, mainly in silver, even though it was basically in production only between 1887 and 1890. The Paymaster shaft house can also be seen slightly above the Silver Belle. It produced about $178,000 in ore.

The Meldrum Tunnel was located about a half mile south of Ironton; it is now caved in. It was the idea of Andrew S. Meldrum, who was one of the original locators of the Yankee Girl Mine. The Meldrum Tunnel was a scheme to permit the use of full-sized narrow gauge railroad cars to travel by tunnel to Telluride, bring back ore, and switch directly onto the tracks of the Silverton Railroad. The bore from each end would have had enough upward grade to drain the great quantities of water that were always encountered deep within the mountains. Ore chutes would have been tunneled into several of the big mines already operating at higher levels than the Meldrum Tunnel, thus providing much cheaper and more convenient transportation of their ores. It was also probable that during the construction of the tunnel, rich veins would be discovered. Andy Meldrum secured his capital from British investors, but the project had just begun when the Boer War broke out in South Africa. Meldrum's money source dried up and his dream ended with only about eight hundred feet of tunnel completed at the Red Mountain end and two thousand feet at Telluride.

The Joker Tunnel was located beside Red Mountain Creek about a mile south of Ironton. Its boardinghouse still stands by the road at the first switchback. The Guston, Gennessee, and Yankee Boy were all nearby mines that were originally developed by way of a shaft (which is vertical) instead of a tunnel or adit (which is horizontal). The mines in this area were following "chimneys" of ore rather than veins, and the chimneys went down to great depths. Eventually the cost of pumping water and hauling ore out of the shafts became very expensive. Aggravating the problem was the fact that the local ores produced acids, when mixed with the ever-present water, which caused the cables and pumps to disintegrate in the mines. The Joker Tunnel was not a mine proper. Its purpose was to drain water and allow economical mining at lower levels. The low-grade ore was loaded directly onto the Silverton Railroad cars, directly outside the tunnel, but unfortunately the ore fell even further in value as the mining went deeper. George Crawford and C. H. Graham spearheaded the Joker Tunnel's construction. It eventually extended forty-eight hundred feet and allowed the Red Mountain mines to operate another decade. The tunnel employed as many as thirty-five men and was used all the way into the 1940s, although 1914 was the end of continuous operations.

The famous Corkscrew Turntable on the Silverton Railroad can be best accessed from the Joker Tunnel. Just hike up the road that goes above the tunnel itself, then simply follow the railroad grade. The turntable was the only one ever built on the mainline of a railroad. It was made necessary by the steep switchbacks at Corkscrew Gulch.

One additional switchback up from the Joker Tunnel, one can find the old road (now Ouray County Road 31) that leads to the other side of the valley to most of

Guston was a jumble of houses, mill and mine structures scattered over the hillside. Author's Collection.

the mines that will be discussed in the following pages. It is the original road to Silverton (the present paved highway was built in the 1940s) and is an interesting side trip (an easy four-wheel-drive route). Or you can stay on Highway 550 and see the mines from a distance. The side road is public, but remember that all of the mines are private property!

Traveling uphill on U.S. 550 past another switchback brings you to a spot where a road leads west up to the Mountain King Mine (also called Betsy and Gold Lion at earlier times). The vein is visible on the surface and runs some six thousand feet down into Ironton Park. It has been prospected and worked off and on for years. Cave-ins were bad at the mine because the walls of the vein were heavily altered (or "flaky). The last big operations were in the mid-1940s.

The side road (County Road 31) crosses Red Mountain Creek and after about a half-mile enters the Guston Mine and then what is left of the ghost town of Guston. On U.S. 550 by looking across the canyon towards Red Mountain, a jumble of buildings (what is left of Guston) are visible below Champion Gulch. The town contained three very famous and large mines—from north to south, the Guston-Robinson, the Yankee Girl, and Genessee-Vanderbilt.

The Guston Mine was located on August 21, 1881, by August Dietlat, Andrew Meldrum, John Robinson, and Albert Lang. At first the mine was operated mainly for its lead, which was needed as a flux in the local smelters. The Guston sold for one hundred thousand dollars in 1883. Development was slow at first, and it was not until 1888 that large-scale production began. Some of the mine's hand-picked ore averaged fifteen thousand ounces of silver and three ounces of gold per ton. The Guston was eventually developed on fourteen levels to a depth of thirteen hundred feet. The

235

profits in 1890 were so good that the Guston paid its stockholders an eighty percent dividend that year. Not one Colorado mine in ten ever paid a cent of dividends to its stockholders—except the Guston and the Yankee Girl! The Guston eventually produced about six million dollars in ore.

Originally, the area around the Guston Mine was called "Missouri City." By the end of 1884 the town of Guston had about twenty buildings including two saloons, two restaurants, and a miner's supply store. Guston also had a post office after January of 1892, but its pride and joy was its church-the only one in the Red Mountain Mining District other than at Ironton. An Englishman, Rev. William Davis, built it in 1891. He did the work almost single-handedly with the miners supplying lumber and money. The church has collapsed, but the steeple can still be found on a hill to the southwest of the Guston Mine. The streets of Guston were steep and snaked around the mountain, one above the other. By 1890 there were 322 persons living in Guston. However, by 1897 the mines had shut down and most of the future development came out through the Joker Tunnel to the northwest.

The Yankee Girl Mine is only about a quarter of a mile to the south of the townsite of Guston and operated concurrently with the Guston Mine for some time. It produced over $8 million during its time (over $6 million in the 1890s alone), and it was far more famous than its sister mines. The Yankee Girl was staked August 19, 1882. John Robinson made the discovery when he picked up a very heavy rock and broke it open to discover it was solid galena (a rich silver-lead ore). The *Red Mountain Pilot* in its March 10, 1883, issue announced:

> The Yankee Girl is the mine that gave Red Mountain the boom. One month after it was located, Silverton parties paid $125,000 for it. This property includes half a dozen locations, two of which are the Orphan Boy and the Robinson, which promise as well as the Yankee Girl. This mine was discovered last August, and up to the time of our visit, had produced $54,288.

The tower-like shaft house of the Yankee Girl is still highly visible. The mine was developed to a depth of 1,050 feet and produced very rich ore of up to three thousand ounces of silver per ton with the added bonus of up to thirty percent copper. By mid-winter of 1882-83 the Yankee Girl was keeping seventy-five mules busy every day, packing its ore to Silverton to be shipped out on the D&RG Railroad. By the end of the winter the mine was reported to have already earned its purchase price. Crofutt reported, "Yankee Girl ore averages $8,000 per ton. Now there are a great many 'bald heads' who would like to hug a Yankee Girl of this character." In 1890 the Yankee Girl paid its stockholders a forty-five percent dividend and shipped $1,352,994 in ore.

The Yankee Girl group was basically five "chimneys" of ore-the Yankee Girl, Orphan Boy, North, West, and South. The mine had its own large powerhouse with two large smokestacks and an extremely attractive boardinghouse. Otto Mears was a part owner of the mine and even went down into the mine several times to watch the mining, so that he could tell the people to whom he gave his famous silver filigree passes that he had seen the silver from the raw material to the finished product. Ore from the Guston and Yankee Girl Mines was so high grade that it was shipped

The Yankee Girl Mine (right) was reaching the height of its production about 1890 while the Guston (left) was just coming into prominence. Courtesy Denver Public Library, Western History Dept. (F9820)

The Idarado Mine was at the peak of its production in the 1950s. Author's Collection.

directly to the smelters without sorting. Unfortunately, the ore lessened in value as it went down, and the Yankee Girl was down to low-grade ore by 1897 when operations ceased. Quite a bit of low-grade ore was just thrown on the dumps and was milled in the 1920s and 1930s at a profit. The Yankee Girl's total production has been estimated as high as twelve million dollars.

The Genessee-Vanderbilt is located about a quarter mile south of the Yankee Girl. If you are traveling the side road, be aware that the road in this area can be a little confusing, but the Genessee is just a little higher than the road to either Red Mountain Town or the Yankee Girl shafthouse. The Genessee is the mine that still has a few steel-covered buildings. Jasper Brown and Adelbert Parsell discovered the Genessee in 1882. As was the case of most mines in the Red Mountain area, it developed slowly. In 1887 its ore ran about thirty dollars a ton in lead and silver. In 1889 the Genessee merged with the Vanderbilt when it was found that the two mines were working the same body of ore. The Vanderbilt was also located in 1882. Production was eventually about one million dollars. The ore was of a lower grade than the Yankee Girl and Guston; however the mine produced considerable silver, gold, lead and copper. It had a forty-five hundred-foot tunnel and a seven hundred-foot shaft with five levels. The mine was intensively worked in the 1940s and off and on even into the 1970s; so a few of its buildings are still standing. About another mile to the south on the side road is Red Mountain Town.

South on Highway 550, the road passes through the Idarado Mine's complex of mine buildings, warehouses, and company houses. There is also a historical overlook at this point. The mine had a checkered career of operation and represents a collection of many of the old claims in the vicinity. The Idarado was first developed as the Hammond Tunnel and then as the Treasury Tunnel Mining and Milling in about 1896. It was hoped that some of the rich gold veins that lay on the Telluride side of the mountain would continue through to this point. A two thousand foot spur was even laid for the Silverton Railroad. For about ten years, operations continued, but then production slumped. The mine was worked only a little until 1937 when San Juan Metals built a new mill, boardinghouse, and other buildings at the site.

At the beginning of World War II, metals were urgently needed for the manufacture of war materials and supplies. The Treasury Tunnel was more or less preempted by the government, which subsidized the drilling of tunnels to tap other mines for base metals such as copper, lead, and zinc. The Sunshine Mining Co. of Grass Valley, California, was given a contract for this rapid tunnel expansion program. In 1944-45, their contract was completed and operation was taken over by the Idarado Mining Co., a newly formed subsidiary of the Newmont Corporation. In 1954 a fire destroyed a large part of the mine's surface buildings, but the company rebuilt. The Idarado Mill operated continuously at Red Mountain with an output of about 800 tons of ore per day. The tailings dump expanded to fill a nearby pond (Malcom Lake), then proceeded down the mountainside to Ironton. In 1956 a new mill with eighteen hundred-ton-a-day capacity was built near Telluride and the mill on the Red Mountain side was shut down. The daily production of ore was then brought out on the Telluride side of the mountain. The mine has more than eighty miles of interconnected tunnels; it connected the old workings of the Tomboy, Ajax, Smuggler, Liberty Bell, Virginius, Pandora, Japan, Barstow,

Argentine, Black Bear, Wheel of Fortune, and many more. The Idarado produced gold, silver, lead, copper, and zinc. It closed in 1979 but is still kept functional by a skeleton crew because ore is still left in the mine. Altogether, the Idarado produced over thirty million dollars in ore.

At a hairpin curve a short distance above the Idarado, a road (private property) leads to the right (west) to the Barstow Mine and Mill and then on to the Greyhound, which has been previously discussed. The Barstow was originally called the Bobtail Mine. It is located in Commodore Gulch, three quarters of a mile northwest of the Idarado. The mine operated from 1895 to 1918. For quite some time, around the turn of the century, it was the only mine shipping ore from the Red Mountain area. It produced slightly less than a million dollars, chiefly in gold, silver, copper, lead, iron, and zinc. It also produced fluorite during World War I. The Barstow has over eleven thousand feet in tunnels. A large boardinghouse and mill were built on the site in the early 1900s. The mill had forty stamps and two Wilfley tables. After the 1930s the Barstow was worked out of the Treasury Tunnel and the Idarado Mine.

Shortly before reaching the top of Red Mountain Pass and a little over a mile above the Barstow cutoff, a small road leads to the left to Red Mountain Town and the National Belle Mine. This is the road that leads directly through the mines that lay on the east side of the valley.

The story of this town is a little complicated because there were two settlements with the name *Red Mountain,* and one of these settlements changed locations and the other merged into it. Red Mountain City was across the divide near the Congress Mine. In fact, the Congress Mine took over its post office when most of the town was moved to Red Mountain Town and the new Otto Mears toll road. Red Mountain City wasn't the only town that "disappeared" in the Red Mountain District. Besides the towns mentioned in this chapter there were Rodgersville, Hudson, Missouri City, and Sweetville. Although they were all touted as up and coming "towns," none was probably more than a cabin and a few tents—all of which made them hard to locate in the 1880s, much less today.

On May 19, 1883, the editor of the *Red Mountain Pilot* complained:

> A large number of our exchanges continue to address us as Red Mountain City and are carried by to Red Mountain Town and frequently lay there a week before the thick-headed postmaster sends them back...All our exchanges will please change their addresses to "Congress, San Juan County" and thus avoid delays.

By March 1883, the original Red Mountain Town (which was located about a quarter of a mile south of its present location) had far outstripped Red Mountain City. It had saloons, doctors, a barber, hotels, mining agents, lawyers, and even a shoe repair and laundry. Yet nearby Rogersville (about a tenth of a mile east of today's remains of Red Mountain Town at the Rogers Mine) got the post office, probably because it was located nearer the toll road. Red Mountain Town then literally picked up and moved in the summer of 1883 to a spot on the toll road itself next to the National Belle Mine. Rogersville and its post office only lasted a few months longer. By May 1883 the new Red Mountain Town had the beginning of a water system and even some fire hydrants. It also had a mayor and a marshal.

A lone wagon makes its way down Red Mountain Town's Main Street about 1895. Author's Collection.

Red Mountain also had its share of prostitutes, whose numbers always increased substantially in the summer. *The Solid Muldoon* of July 24, 1885, announced, "The rainy season is here. A profusion of flowers and soiled doves hereabouts." And the next week the same paper noted that the Red Mountain "dance hall is kept in full blast night and day—some days. Whoop her up boys, the season will not last long." In the winter the prostitutes returned to Ouray or Silverton. It was not an easy life for them. For example, on September 15, 1886, Willard Donnehauer assaulted Lulu Waters, better known as Indian Lou, and pounded her over the head with a gun while another man watched. Both men were arrested and charged with assault with intent to murder. As more and more men brought their wives and children to the camp, the prostitutes dwindled in numbers.

By 1885 Red Mountain Town had a population of 598 with its own telephone office, two newspapers, schoolhouse, post office (it boasted inaccurately that it was "the highest in the world"), and many saloons. Rev. J. J. Gibbons reported in 1891 that:

> The lights never went out in the camp, unless when coal oil failed, or a stray cowboy shot up the town. The men worked night and day, shift and shift about and the people were happy. The gambling halls never closed. The restaurants did a profitable business, and no one could lay his weary bones on a bed for less than a dollar.

The town was plagued by fires—one major one on August 13, 1892, and another on June 13, 1895. Each destroyed over half the town despite the fact the town had its own waterworks and a very good volunteer fire department. By 1893 Red Mountain's population had dropped to about four hundred and in 1894 to two hundred. By 1898 the winter population was only twelve, and in 1900 thirty people were there to be counted by the census, but they lived at Red Mountain only in the summer. Much of what remained of the town was destroyed in a forest fire in 1937. The jailhouse and a few foundations are about all that is left. The jail was made of tons of solid wood (two by sixes on top of each other) and was built on solid bedrock so that no one could dig out.

The National Belle Mine was located in January 1883 and ended up within the city limits of Red Mountain on the big knob at the end of town. It was an early producer of soft carbonates, galena, and gray copper which were not very rich (seventy-five dollars per ton or so) but which could be mined very easily and cheaply. The mine would eventually produce over $2 million in ore. Six months after the mine was located, the owners were offered $160,000 for the mine. It was in July 1883 that the big excitement happened. Workers discovered an immense cavern that was like a treasure cave—filled with pockets of gold and silver galena as well as chlorides and carbonates. Travel writer Ernest Ingersoll reported:

> A workman broke through the walls into a cavity. Hollow echoes came back from the blows of his pick, and stones thrown were heard to roll a long distance. Taking a candle, one of the men descended and found himself in an immense natural chamber, the flickering rays of the light showing him the vaulted roof far above, seamed with bright streaks of galena and interspersed with masses of soft

Red Mountain's residential section sat south of the National Belle Mine that dominates the town. Moore Photo. Courtesy Ouray County Historical Society. (X13131)

carbonates, chlorides and pure white talc. On different sides of this remarkable chamber were small openings leading to other rooms or chambers, showing the same rich formation. Returning from this brief reconnaissance a party began a regular exploration. They crept through the opening into an immense natural tunnel running above and across the route of their working drift for a hundred feet or more, in which they clambered over great bowlders (sic) of pure galena, and mounds of soft gray carbonates, while the walls and roof showed themselves a solid mass of chloride carbonate ores of silver. Returning to the starting point they passed through another narrow tunnel of solid and glittering galena for a distance of forty feet, and found indications of other large passages and chambers beyond...It would seem as though Nature had gathered her choice treasures from her inexhaustible storehouse, and wrought these tunnels, natural stopping places and chambers, studded with glittering crystals and bright mineral to dazzle the eyes of man in after ages, and lure him on to other treasures hidden deeper in the bowels of the earth.

People came from miles around to gawk at the treasure cave. The main cave was two hundred feet long and one hundred feet wide. Within a week an even larger mineral cave was discovered. The *Red Mountain Pilot* reported (7/21/83):

No where else in the world is there such a large body of mineral as there is in the National Belle and no one knows the extent of the mineral. There is at least a million tons in sight and no one can estimate the value. ...One of the best features about this rich discovery is that the present owners were the original locators and they will reap the fruits of their toil instead of some tin-horn capitalist.

The mine became one of the best known in the country and eventually produced almost $9 million in ore. In 1885 Crofutt reported that "the National Belle is another mine that pays those who wring her, big!" But another very important factor was the low production costs of the National Belle. It was just a few feet away from the Silverton Railroad and its ore was very soft and easy to mine—often just shoveled into sacks with no blasting or sorting needed. The shaft house of the National Belle still stands above the dump and loading chutes. At first the mine produced huge quantities of up to sixty percent lead, but then as production grew deeper, copper of up to forty percent was found. In 1897 the mine closed after being developed to a depth of 450 feet.

There are several back roads in the vicinity of Red Mountain Town. One leads up the mountain to the east past the Hudson Mine (which was also called Hudson Town at one point) and on up to several newer mine workings. Another road leads to the south about a mile and comes out at the Longfellow Mine.

Back on Highway 550 from the Red Mountain Town road, and a little further south, a pack trail leads north and west up to Ptarmigan Lake, the Senator Beck Mine, and finally to Imogene Pass. This used to be called "The High Line Trail." Hundreds of tame ptarmigan could be found along the way. Today, both sides of the highway provide access to good cross-country skiing. Near this point is the Black Bear jeep road. The road is named for the Black Bear Mine in Ingram Basin on the other side of the pass near Telluride.

If you are going to travel over Black Bear toward Telluride from Red Mountain, it is two-way traffic until the switchbacks going down into Telluride. From this point on it is one way down to Bridal Veil Falls. The beginning of the road is a fairly easy jeep road for the first six miles, but the balance is an extremely dangerous section of road (one of the hardest in the San Juans) because of the tightness of the switchbacks, which require vehicles to back up, sometimes on loose shale. The total trip takes two to three hours (whether going all the way to Telluride or turning around and coming back to U.S. 550) and is about twelve miles. People have been killed on the Black Bear switchbacks—be very careful! This is the road that had a sign for many years that read, "You don't have to be crazy to drive this road, but it helps!"

To travel Black Bear, at about one steep mile from U.S. 550, stay on the main road to the right. The road that leads to the left goes to Porphyry Basin and the Bullion King Mine. At a little over two miles from U.S. 550 (and above timberline) the road levels out in a hilly area. The top of the pass is about three and a quarter miles from U.S. 550 and is at 12,840 feet. Trico Peak is to the right (north). It is so named because it is the boundary line between three counties. As you approach the ridgeline the easier route is to the left. After you cross the ridge the entire San Miguel Valley comes into view. At about four miles go straight. The road to the left goes out to the Andres Mine. Ingram Creek and turquoise-colored Ingram Lake (also called Constantine) are on your left. The mine on your right, above the road, is the Black Bear Mine. It was located in 1878 and originally worked by Finnish miners. Unlike many of the local mines it was mainly owned by local Telluride investors and therefore wasn't as subject to labor strikes in the early 1900s as some of the other Telluride mines. It had a small crusher, a boardinghouse, and a tram that ran to the Smuggler-Union Red Mill in Pandora, but there never seemed to be enough capital to do any large-scale development. There was an angle station (it still stands with its rounded top) at Ingram Falls, and also a mill but it never really worked correctly. During the 1920s some mill equipment was moved to the mine entrance, and a bigger boarding house was built. An avalanche hit the large building (it held 150 miners) in 1923 and injured three people and killed two. The mine was basically abandoned at this time. Although the Black Bear had some very valuable ore, it was never very profitable until it was consolidated into the Idarado and worked underground for its gold. The road was reopened in 1956 by local jeepers.

About two and one-half miles past the ridge are some of the hardest parts of the road. Your passengers may want to get out and walk at this point if they aren't doing so already. The switchbacks get easier as you go down and eventually you come to the western two-way section of the road, which is by the gate to the powerhouse at Bridal Veil Falls power house. The falls all but disappeared when the Smuggler-Union Mill was working and piped the water for power to its mill. The Pandora Mill is located shortly before reaching the pavement. Telluride is then a couple of miles down the road.

About a mile up above Bridal Veil is Gray's Basin. It is blocked off from vehicular traffic but holds the Mayflower Mine, which was worked in the 1890s and 1900s—its two-and-a-half mile tram was constructed in its later years and was the longest in the area. It had a twenty-stamp mill near Telluride by the RGS tracks. The Lewis Mine is below the Mayflower in Bridal Veil Basin. Its ore averaged sixty dollars a ton in 1882. A mill was constructed at the mine, and it operated off and on into the 1950s. Another longer trail comes to the mine from the Million Dollar Highway. The Lewis Mill is probably in the best condition of any mill in the San Juans, except the restored Sound Democrat Mill in Placer Gulch.

Around the valley to the west (but also accessible only to hikers and not to vehicles) is the Bear Creek Mine and Mill, which are located on the creek with the same name. The mill is about two miles out of Telluride and was built in 1893. The mine is high up in the cliffs to the east, and it was connected by cables to the mill. At one time the mill had 120 stamps. It was shut down about 1914. Now only rows of concrete pedestals mark the location.

The Nellie Mine was another mile south on the west side of Bear Creek. It is perhaps the most famous for the early day photographs of the mule train that brought the tram cable to the mine. The Nellie produced about $53,000 in 1884, then shut down. The mine opened again and produced thirty million dollars a month for two months in 1898 but then stopped shipping ore because of litigation. The Mammoth, Silver Cliff, Maryland, and Talley Thomas Mines are also on Bear Creek. The whole area was described in the 1880s as Telluride's "Gold Belt," but the gold-rich Liberty Bell, Smuggler-Union, and Tomboy Mines had yet to make their rich strikes.

Back on U.S. 550, Red Mountain Pass summit is the county line between Ouray and San Juan and is also the division between the San Juan and Uncompahgre National Forests and the Uncompahgre and Animas River watersheds. U.S. 550 basically follows the old grade of the Silverton Railroad from this point south. A small community at the crest of the pass was originally named "Sheridan Junction" and then later "Summit" by the Silverton Railroad because it was the highest point on their line. Several years earlier this was also the location of the town of Congress. The town of Congress was built in the dead of winter during the original Red Mountain rush. The cabins were laid directly on the hardened snow, and they tipped every which direction when the snow melted in the spring (summer at this altitude). Its population rose to about 130 during its prime (about 1885), but it fell back to only a few people by 1887. In 1884 the town had a population of about fifty including several stores, a post office, and a hotel.

The Silverton Railroad was Otto Mears's first rail line, which he started in 1887. Construction started in Silverton that summer and reached Chattanooga by October. In the spring of 1888 construction crews made their way up the famous Horseshoe Curve above Chattanooga. The tracks made a 200 degree turn and maintained a constant five percent grade for one and three-quarter miles. On September 19, 1888, the railroad reached Red Mountain Town, where the depot was constructed inside a wye and a short spur was constructed to the National Belle Mine. By the end of 1888 the line was near Corkscrew Gulch, and spurs were created to all the large mines along the route. By the middle of 1889 the turntable was in operation (although it wasn't covered until the first snows made it obvious that it would be necessary). By late summer the railroad was in Ironton. The Silverton Railroad really opened up the Red Mountain Mining District as many low-grade ores could be shipped economically. The Silverton Railroad was nicknamed "The Rainbow Route." The wealth of scenery along the way as well as the pots of gold (unfortunately mainly silver) at the mines in the Red Mountain Mining District were good reasons for the name. The Silverton Railroad made twice the normal rate of return for a Colorado railroad—enough profit to finance a great deal of

The Longfellow Mine is perched to the east of U.S. 550 near the top of Red Mountain Pass.
Author's Collection.

Bear Mountain looms at the end of the valley behind the settlement of Chattanooga. Courtesy
Colorado Historical Society. (J13830)

the construction for Mears's Rio Grande Southern Railroad that ran from Ridgway to Durango. One reason for the railroad's good economic showing was Mears's incredible timing, which matched the building of the Silverton Railroad with the high prices of the Sherman Silver Purchase Act of 1890.

Shortly after the summit of Red Mountain Pass a road to the left leads to the Longfellow, St. Paul, and Congress Mines. The Longfellow is right alongside U.S. 550 and was worked until the 1950s. About a quarter of a mile down the road is a weather station that transmits all its information by satellite. A half-mile further is the St. Paul Mine, which is now a cross-country ski resort. The Congress Mine is two-tenths of a mile further. It was partly owned by Thomas Walsh at the same time he was starting the famous Camp Bird Mine. The Congress was one of the first mines discovered at Red Mountain. By July 1881, the Congress had been sold for twenty-one thousand, and then it was sold again. Its ore averaged about sixty dollars per ton, which was well worth shipping after the arrival of the D&RG Railroad in Silverton in July 1882. In 1883 the mine produced ore worth $220,000-a good sum for those days.

Above the Congress Mine was the site of Red Mountain City. A few fallen down buildings still remain. Although Red Mountain City was initially "the town" on Red Mountain, by April 1883 it was already declining. By 1884, the town had basically disappeared.

There is a locked gate above Red Mountain City, but a jeep road leads by the St. Paul and continues on for five miles to the Brooklyn Mine and then back down to U.S. 550 near the Ophir Pass road. This road is not recommended when wet, but it goes high above tree line and has some beautiful views.

Back on U.S. 550, Porphery Basin, or the Bullion King Road, is the well-traveled dirt road that leaves to the right about a mile and a half after starting down from the level spot at the top of Red Mountain Pass. It is also accessible off the Black Bear Road, which ties in from the north about six-tenths of a mile up the road. About a mile further is the Bullion King Mill, and about a half-mile hike above the mill is Bullion King Lake (actually two lakes). Except for a narrow section near the top it is a fairly easy drive, the views are spectacular, and the flowers are usually in great profusion.

As Highway 550 descends, the view down Mineral Creek includes the headframe and dump of the Silver Ledge Mine, located in 1883. On March 3, 1883, the *Red Mountain Pilot* reported that Chattanooga was thrown into a fever by the discovery of the mine and that it was the largest body of ore the editor had seen. It didn't turn out that way, but there was a good-sized settlement at the mine for many years, as well as another settlement at the Silver Ledge Mill down below in the Chattanooga Valley. At the mill the ore could be dumped directly into Silverton Railroad cars. For all of the nineteenth century, mills charged a penalty for ore with zinc content because it was hard to separate from the other valuable minerals. In 1904 the Silver Ledge began the first economic recovery of zinc. The mine and mill operated off and on for decades and even had a post office from September 1904 to March 1905, but the mine shut down after the mill burned in 1919.

The Million Dollar Highway then enters the Chattanooga Loop, which was another of many engineering feats of Charles W. Gibbs. The railroad curve at Chattanooga Loop (also called Muleshoe Curve) was two hundred degrees, encompassing a grade of 550 feet in a quarter-mile. Mill Creek is to the west. The 1880s saw the first significant mining at Mill Creek. The Hoosier Bay was one of the mines in the gulch at the bend in the loop. It struck rich gold ore in 1893 just as silver was playing out and sparked a gold rush to the area that resulted in fifty-two new claims being filed. Unfortunately, the gold soon played out.

In the valley beyond the curve are the remains of the town of Chattanooga (called "Silver Crown Camp" before the name was changed to Chattanooga). The view ahead toward Silverton is dominated by Bear Mountain, so named because the trees on its hillside create a profile of a bear with upraised paws holding a honeycomb. (Can you make it out? Look for the "V" formed by the trees in the distance.)

Although a few people lived in the area as early as 1878, Chattanooga was founded in March 1883. Within thirty days of its founding (April 3, 1883), Chattanooga had a post office, which remained open until 1894. By the middle of May of the same year, the town contained seventy-five buildings in various stages of construction including a hotel, two general stores, two bakeries, two drug stores, two meat markets, a blacksmith shop, several restaurants, and numerous saloons. Chattanooga was mainly a supply town; the wagon road from Silverton ended at this point and supplies bound for Red Mountain were transferred to mules or burros. The ore came back the same way. In 1884 Mears completed his toll road from Silverton to Red Mountain and Chattanooga's usefulness ended. In 1885 its population dropped to

William H. Jackson took this photo of what was left of Chattanooga after an avalanche hit it. Courtesy Colorado Historical Society. (WHJ32580)

The Fischer Brewery (in the foreground) and the Walsh Smelter (in the background) in 1891. Courtesy Denver Public Library, Western History Dept. (X1739)

about sixty. Then the town was partially destroyed in 1888 by a snowslide, and a fire in 1892 finished off most of what remained. The Silverton Railroad further doomed the town when its construction train passed through Chattanooga in 1888. By 1890 the town had only two residents—a saloonkeeper and a widow who took in laundry. A few people moved back to the town when the Silver Ledge Mine built a large mill there in 1902, and there were enough children that a school was open in Chattanooga that year.

U.S. 550 follows Mineral Creek past its confluence with Mill Creek, and after about a mile and a quarter reaches a turnoff to the east. This road leads two miles up the mountain to the Brooklyn Mine, which was operated as recently as the 1980s and is still in decent shape. The Brooklyn was operated mainly for its gold. The road should not be attempted if wet.

Continuing along U.S. 550 about three-quarters of a mile, the turnoff to the west (right) over Ophir Pass is the location of Burro Bridge (an early-day settlement of a dozen cabins and a few barns and stables). This is the start of the four-wheel-drive road that leads over Ophir Pass to the towns of Old Ophir and New Ophir on the Howard Fork of the San Miguel River. The Ophir Pass Road was originally a toll road, opened in 1881 to carry ore from Ophir to the mills in Silverton. It is now one of the easier four-wheel-drive roads in the area and is described in detail in another chapter.

Then after three more miles, the cutoff to South Mineral Creek is to the right (west). This was the original route into Silverton from Durango, but now the road dead-ends after seven miles and the pass can be traveled only on foot. There are several good camp sites and usually excellent fishing along the creek, as well as some beautiful views. Follow San Juan County Road 7 (Forest Service 585) west for almost four miles to Forest Service Road 815, which forks to the right (south) and goes to Clear Lake. The road has a series of switchbacks that lead four and a half miles to the lake (which sometimes has good fishing). Back on County Road 7, it's only a quarter of a mile to the South Mineral Campground (a fee campground), which has good camping, fishing, and toilet facilities. The trail to Ice Lake also starts here (another branch is at the first switchback up the Clear Lake Road). A jeep road leads southwest out of the campground for two miles to the Bandora Mine and then a trail continues on the original route into Silverton from Durango and another fork goes on to Rico (neither are passable by vehicles). There was a small settlement called Bandora near the mine in the 1880s. Gold was found in the area as early as 1882, and the Bandora Mine was producing well in the 1890s. Its ore was high grade but was very hard to smelt.

As you come into Silverton on U.S. 550 the ruins of the large North Star Mill are across the creek at the base of Sultan Mountain. The North Star (not to be confused with the North Star Mine only a few miles away in Arrastra Gulch) was discovered in 1876 and was producing well by 1880. By 1882, with the arrival of the D&RG Railroad, it was shipping ore with seventy to eighty ounces of silver and thirty-five percent lead. That year the North Star took out several huge specimens that were almost pure gray copper. In 1883 its ore was assayed at up to one hundred ounces of silver, a half ounce of gold, and forty percent lead. Thomas B. Comstock built the Comstock Sampling Works near the mine in 1883. It was a concentrating mill and sampling works, which employed seventy men. Comstock sold the mill to the mine in 1886, and it was known thereafter as the North Star Mill.

Continuing further down U.S. 550, the black dump to the left just outside Silverton is from the Walsh Smelter. Originally called the Martha Rose Smelter, it was built in just 104 days in 1882, however it was open only three days and then shut down until 1894 when Thomas Walsh bought the smelter. The jet-black, flint-like material is the residue from the smelting process. Walsh was looking for flux for this smelter when he discovered the famous Camp Bird Mine.

Across the road from the dump was the famous Fischer Brewery. Fisher at one time advertised his beer as "a tonic not an intoxicant." Some of the stone from the brewery was used to build the Christ of the Mines Shrine above Silverton. A great part of the brewery's fame came from a spigot that was outside the door where any visitor was welcome to drink their beer free of charge. Unfortunately, nothing remains today!

A few more miles down the road we reach Silverton—the place where it all started and the beginning of our adventures. To travel all the roads and side roads, to hike all the trails, and to explore that part of the country without trails in the San Juans would take a person a lifetime. I hope that this book has covered all the high points and given you an appreciation for the awesome beauty and bountiful history of this very special place. No wonder one of the favorite nicknames for the San Juans is "God's Country." If God were to choose a home on this earth, there is no doubt that the San Juans would be one of his first choices. Enjoy!

INDEX

NOTES

NOTES

NOTES

NOTES

NOTES

NOTES

NOTES

NOTES

NOTES

NOTES

CPSIA information can be obtained
at www.ICGtesting.com
Printed in the USA
LVHW052258060621
689421LV00008B/60